From WPANs to Personal Networks

Technologies and Applications

For a complete listing of the *Artech House Universal Personal Communications Series,* turn to the back of this book.

From WPANs to Personal Networks

Technologies and Applications

Ramjee Prasad

Luc Deneire

ARTECH
HOUSE

BOSTON | LONDON
artechhouse.com

Library of Congress Cataloging-in-Publication Data
A catalog record of this book is available from the U.S. Library of Congress.

British Library Cataloguing in Publication Data
Prasad, Ramjee
  From WPANs to personal networks: technologies and
  applications. — (Artech House universal personal communications series)
  1. Personal communication service systems   2. Wireless LANs
  3. Bluetooth technology
  I. Title   II. Deneire, Luc
  621.3'821

  ISBN-10: 1-58053-826-6

Cover design by Igor Valdman

International Standard Book Number: 1-58053-826-6

10 9 8 7 6 5 43 2 1

*To my wife Jyoti, to our daughter Neeli, to our sons Anand and Rajeev,
and to our grandchildren Sneha, Ruchika, and Akash*

—Ramjee Prasad

To my wife Isabelle and to our sons Antonin and Basile,

*To all my teachers, amongst whom my parents were the first, they taught
me to be fierce with work and gentle with people.*

*To professors Spronckard Fawe, from University of Liège, Belgium,
who introduced me to science and technology, and above all, by their
patience and friendliness, taught me to place the person at
the center of my thoughts.*

—Luc Deneire

Contents

Preface

यदृच्छालाभसन्तुष्टो द्वन्द्वातीतो विमत्सरः ।
समः सिद्धावसिद्धौ च कृत्वापि न निबध्यते ॥२२॥

yadṛcchā-lābha-santuṣṭo
dvandvātīto vimatsaraḥ
samaḥ siddhāv asiddhau ca
kṛtvāpi na nibadhyate

Gita (4.22).
He who is satisfied with gain which comes of its own accord, who is free from
duality and does not envy, who is steady in both success and failure,
is never entangled, although performing actions.

Personal area networks (PANs), a concept introduced by Zimmerman in the
mid-1990s, are meant to enable (both data and multimedia) communications
between people at conversational distances (i.e., about 10 meters). These PANs
transform persons in smart devices, which can exchange their digital intelligence
in a seamless way. The European Commission Sixth Framework Integrated Project, "My Personal Adaptive Global NET (MAGNET)," is leading the technology in this direction.

The threefold objective of this book is to give an overview of the state of
the art on wireless PANs (WPANs), to describe the motivations and projects that
are behind the emergence of WPANs, and to conjecture on the future of
WPANs. The extension of WPANs is towards personal networks (PNs), which
lead the path towards the fourth generation (4G).

The prehistory of (digital) wireless networks goes back to the 1980s, when
voice communications was brought to the nomadic person, albeit through expensive operators, followed by data communications (with wireless local area
networks) through unlicensed bands. When Ericcson imagined to make connection cables disappear and Zimmerman introduced the PAN concept, they
were at the eve of a new era: the nomadic person became a communication device in and of himself, and he began asking for worldwide accepted standards to
exchange his data securely without tedious manipulations.

Part 1, along with the introductory chapter, will elaborate on the initial motivations and challenges of WPANs and will give a historical and technical overview of WPANs. It will detail the architecture of its two most well-known technologies, Bluetooth and IEEE 802.15.

The success of the WPAN paradigm is based on a strong consensus that new technologies should be centered on the user, improving the quality of life and adapting to the individual. Along with this concept come the ideas of invisibility, automatic connection, service discovery, security, and computing, all within the personal operating space (POS). This POS, a new concept in communications, is what makes WPAN a revolution in the wireless world. The POS is tethered to an individual and is created by the communication between its devices. Furthermore, it can communicate to other POSs and with the outer world, hence playing the role of universal access interface from and to the networked user.

Underpinning this POS, and nurturing the WPAN concept and technical possibilities, are a huge number of projects, from sensor networks to those involving ambient intelligence. These projects mainly deal with three main aspects: wireless ad hoc networks, low power, and security issues.

Part 2 will elaborate on the motivations behind new WPAN approaches and describe, in the light of these motivations, various research activities relevant to the three main aspects of WPANs. Security is also introduced as one of the many technology challenges.

WPANs were initially meant to be present in license-free bands, thus escaping the classic operators. Still, to be commercially viable, WPANs have to insert themselves in the current wireless networking landscape. At the same time, one of the important direction in future 4G networks is the integration of the personal dimension, including WPANs.

Part 3 will discuss the relations between WPANs and 4G networks, as well as the impact of the vertical and horizontal (in terms of current standards, read Bluetooth versus 802.15) approaches on personal network design. Taking the integration of these heterogeneous networks into account, as well as the results of related research projects, we will draw the lines of what could be the future extension of WPANs towards PNs and applications.

Figure P.1 illustrates the coverage of the book.

This book will help to solve many problems encountered in the research and development of WPANs, PNs, and 4G. We have tried our best to make each chapter comprehensive and we cannot claim that this book is without error. Therefore, we welcome any comments to improve the text and correct any errors to deneire@unice.fr.

We acknowledge Mrs. Junko Prasad for preparing the typescript of the book.

RAMJEE PRASAD
LUC DENEIRE

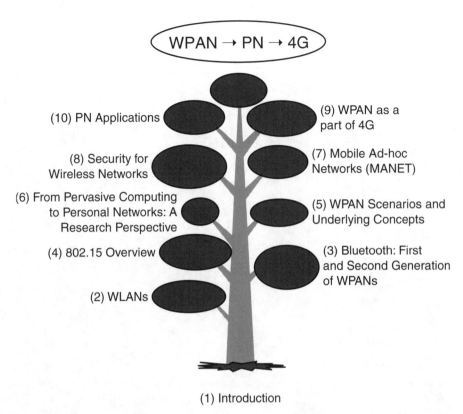

Figure P.1 Illustration of the coverage of the book. The numbers indicate the chapters of the book.

1

Introduction

Danish King Harald Blatand—for whom, a thousand years later, Bluetooth for wireless personal area networks was named—is known for uniting parts of Sweden, Denmark, and Norway. Uniting computers, mobile phones, and personal devices is the goal of wireless personal area networks (WPANs), which are meant to become a major part of future mobile communication networks and the fourth generation (4G). This introduction provides an abstract view of what a WPAN is, or should look like.

The personal area network (PAN) is a network for *you*, for *you and me*, and for *you and the outer world*. It is based on a layered architecture where different layers cover the specific types of connectivity (see Figures 1.1 to 1.3) [1–15].

This connectivity is enabled through the incorporation of different networking functionalities into the different devices. So, for the stand-alone PAN, the person is able to address the devices within his personal space independently of the surrounding networks. For direct communication between two persons (i.e., their PANs), the bridging functionality is incorporated into each PAN. For communication through external networks, a PAN implements routing and/or gateway functionalities.

Layer-oriented scalable architecture supports the functionalities and protocols of the first three layers and provides the capability to communicate with the external world through higher layer connectivity. It provides the appropriate middleware structures and consists of a well-defined protocol stack, with identified information transfer through appropriate interfaces.

The PAN can use various access technologies, calling for reconfiguration. Moreover, according to the applications, PAN systems provide automatic service and resource discovery, provide QoS (e.g., for multimedia applications), and are scalable in terms of network size.

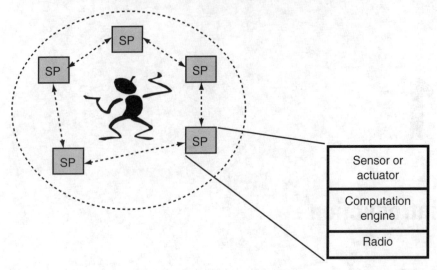

Figure 1.1 PAN is for *you*. A PAN constructs a personal sphere of smart peripherals (SP).

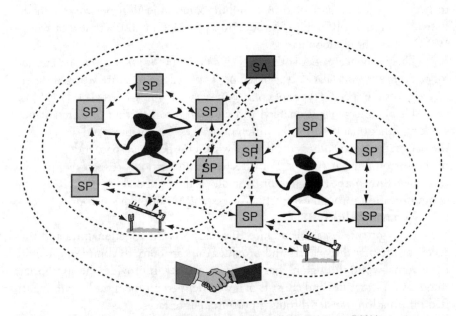

Figure 1.2 PAN is for *you and me*. When people and appliances meet, a PAN becomes a dynamic distributed application platform where gatekeepers are needed.

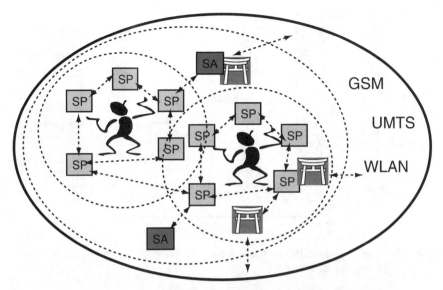

Figure 1.3 PAN is for *you, me, and the outer world*. Extending your reach requires a multimedia gateway as well as a distributed resource control with quality of service (QoS). (GSM: Global System for Mobile Communication; UMTS: Universal Mobile Telecommunications System; WLAN: wireless local area network.)

PAN invisibility is essential to the user, and so, PAN devices are able to adapt themselves automatically to the environment and, for example, can download the appropriate applications and access techniques automatically.

Frequency planning and coexistence with existing systems is important for designing novel PANs. PAN-oriented applications mostly use unlicensed frequency bands. For higher data rates, the 5-GHz frequency band, and possibly 60 GHz, can be used (Figure 1.4), but the advent of ultra-wideband (UWB) access techniques already offers data rates of up to 1 Gbps for very short ranges.

1.1 Possible Devices

Different services and applications in a PAN lead to different terminal functionalities and capabilities. Some devices, such as simple personal sensors, are very cheap and incorporate limited functionalities. Others incorporate advanced networking and computational functionalities, which make them more costly. Scalability of the following elements is crucial:

- Functionality and complexity;
- Price and power consumption;

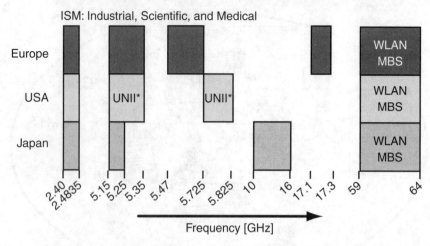

Figure 1.4 Frequency bands. (MBS: Mobile Broadband System; UNII: Unlicensed National Information Infrastructure.)

- Data rates;
- Trustworthiness;
- Supporting interfaces.

The most capable devices incorporate multimode functionalities that enable them to access to multiple networks.

Some of the devices are wearable or can be attached to the person (i.e., sensors); some are stationary or associated temporarily to the personal space (i.e., environmental sensors, printers, information desks). A set of possible devices for PAN applications is presented in Figure 1.5.

1.2   PAN Challenges and Open Issues

The development of PAN architectures and devices creates many challenges for the scientific and industrial communities. Indeed, PAN builds up applications on a dynamic distributed platform and provides distributed resource control with QoS. It relies on low power and low cost radios and a variety of reconfigurable terminals. The PAN disappears into the living environment, allowing seamless network connectivity and secure communications. To fulfill these goals, PAN requires the following:

- Low power, low cost radio integration;
- Definition of possible physical layers and access techniques;

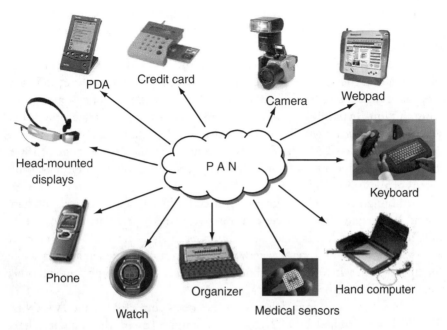

Figure 1.5 Possible devices for PAN applications. (PDA: personal digital assistant.)

- Ad hoc networking;
- Middleware architecture;
- Security (different security techniques, gate-keeping functionalities);
- Overall system concept;
- Human aspects.

These challenges are studied in the various projects that are described in Chapters 6 through 8.

Standardization activities regarding PAN are ongoing within the IEEE 802.15 groups [1]. This organization tries to standardize the overall system concept as well as point out the general requirements for PAN as guideposts for researchers and industry to bring PAN into everyday life. Chapter 4 provides an overview of 802.15 activities.

1.3 WLANs versus PANs

Regarding the rapid development of WLAN standards in recent years, as well as some of the target WLAN applications, a natural question arises: why research PANs when there is already a well-traced line of progress for WLANs? WLANs can also afford wireless connectivity to proximate portable computing devices, which is an initial drive for designing PANs. However, there are some important differences between WLANs and PANs.

Wireless PAN technologies emphasize low cost and low power consumption, usually at the expense of range and peak speed. Wireless LAN technologies emphasize higher peak speed and longer range at the expense of cost and power consumption. Typically, wireless LANs provide links from portable laptops to a wired LAN via *access points*. To date, IEEE 802.11b,g,a has gained acceptance rapidly as a wireless LAN standard. It has a nominal open-space range of 100m and a peak over-the-air speed of 54 Mbps. Users can expect maximum available speeds of about 30 Mbps.

Although each technology is optimized for its target applications, no hard boundary separates how devices can use wireless PAN and wireless LAN technologies. In particular, as Figure 1.6 shows, both could serve as a data or voice access medium to the Internet, with WLAN technologies generally best suited for laptops, and wireless PAN technologies best suited for cell phones and other small portable electronics.

There has been a long debate over the coexistence of PAN and WLAN [2, 3], which was studied by 802.15.2, for example. Lately, the standards have evolved in such a way that coexistence is taken into account from the beginning of the design process. Moreover, the advent of ultra-wideband communications has lead to fierce debates on interference between UWB and narrowband links, leading to stringent UWB regulations (see Chapter 4).

1.4   Personal Networks Concept

The concept of the personal network (PN) goes beyond the commonly accepted concept of a PAN [17]. The latter refers to a space of small coverage around the person where ad hoc communication occurs. This is also referred to as a personal operating space (POS). PNs extend the local scope of PANs by addressing virtual personal environments that span a variety of infrastructures (as well as ad hoc networks). Even though we have described the PAN view as addressing the problem of communication between *you and the outer world*, PN extends the PAN concept even further, as the POS can be distributed all over the world.

Let us first describe some potential scenarios involving PNs.

- *A health-monitoring application.* A disabled or elderly person has a PAN incorporating sensing devices linked to a health-monitoring server at home. As this person moves away from home to another location, the server stays connected at all times to the sensing devices in a PN, which

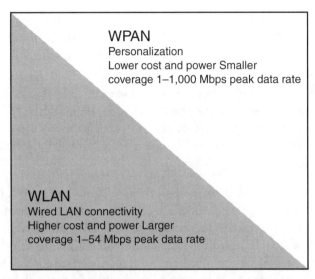

WPAN
Personalization
Lower cost and power Smaller
coverage 1–1,000 Mbps peak data rate

WLAN
Wired LAN connectivity
Higher cost and power Larger
coverage 1–54 Mbps peak data rate

Figure 1.6 Characteristics of WLANs and WPANs.

is formed by linking the PAN-connected devices via, for example, a third generation (3G) network and the Internet to the home network, where the health-monitoring server resides.

- *Walking through a building.* As a woman walks through a smart building from room to room, a PN accompanies her. It interacts with the building functions and controls the lighting, enables access to restricted areas, and activates building devices. The PN could, for instance, incorporate a large wall-mounted display where the woman can view an incoming video stream directed to her, one that cannot be displayed properly on her PDA.

- *Extending the business environment from the office to the car.* A man leaves his office and enters his car. A PAN is established that incorporates a number of car information accessories (via the on-board car network) so that he can listen to his corporate e-mail text read by a computer, as well as dictate and send replies. This could be realized, for instance, by linking up and temporarily extending the person's PAN containing a 3G-enabled PDA with on-board speakers, microphones, and a voice-recognition and synthesis system.

- *A telepresence scenario.* One or more video cameras and high-quality displays surround a person in the office and at home. These devices are incorporated, automatically and invisibly, into this person's PN as he enters the office or sits down on a couch in his living room. They allow him to start up a telepresence session via, for example, his PDA, through which he can have a virtual meeting with other people for business or for social occasions.

- Alternately a person on the move could carry around some high-quality portable wireless screens and cameras, which could be spread around to emulate the presence of remote participants in a session. Again, this would involve the establishment of a PN involving local and remote devices.

Let us envisage how these example scenarios could happen. An individual owns a PAN, consisting of networked personal devices in close vicinity (e.g., attached to the body or carried in a briefcase). This PAN is able to determine its context (e.g., where it is), and interact and link up with devices in the environment or with remote devices in order to temporarily create a PN. This PN provides the functionality (e.g., office functions in the car) the individual wants at that very moment and in that particular context.

Referring to the multisphere model proposed in the *WWRF Book of Visions* [18], a PN runs across the spheres defined around a person: starting from the PAN sphere and ranging via the immediate environment, the instant partners, and the radio access and the interconnectivity spheres to the cyber world sphere. It reaches out to whatever resources or partners are needed to support and enhance a person's private and professional activities. These resources and partners are not necessarily in the immediate geographic vicinity of the person.

PNs are very much centered on a person and his or her needs. They will be dynamic in composition, configuration, and connectivity depending on the time, the place and circumstances, the resources required, and the partners one wants to interact with.

We envision a PN to have a core PAN consisting of devices that a person carries with him most of the time (e.g., a combined PDA-cellular phone). This core PAN will, if its owner desires, look out continuously for what the electronic environment has to offer. Alternatively, if a user values privacy or isolation under given circumstances, his core PAN will isolate itself from this environment. The core PAN is extended, on demand and in an ad hoc fashion (driven by the opportunity and the applications), with personal resources or resources belonging to others (organizations or people). The resources that can become part of a PAN will be very diverse—for example, computers, PDAs, phones, headsets, displays, Internet-enabled appliances, sensors, and actuators. There are many more devices with communicating and processing capabilities that will emerge in the coming years (this is the main rationale for the IPv6 protocol). These resources can be private or may have to be shared with other people. They may be free or may involve some cost for their usage.

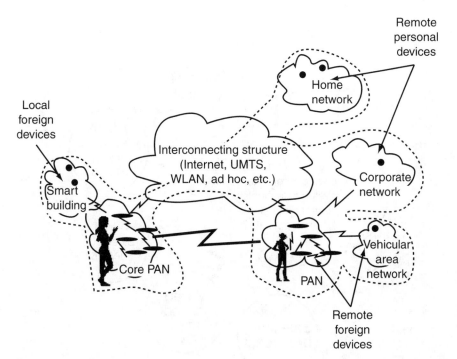

Figure 1.7 Personal network.

The extension of the PAN with remote devices will physically be made via infrastructure networks (e.g., the Internet, an organization's intranet, or via a PAN belonging to another person, a vehicle area network, or a home network). Figure 1.7 illustrates the concept of a personal network.

An important new element suggested by the figure is that the composition, organization, and topology of a PN are determined by its context. By this we mean that the geographical location of a person, the time of day, the electronic environment, and the explicit or implicit wishes to use particular services determine which devices and network elements will be incorporated in a PN.

1.5 Fourth Generation

The concept of 4G is discussed in detail in Chapter 9. Convergence is really what 4G is about. From the wireless perspective, cellular systems (2G, 3G), WLANs, WPANs, satellite communications, and broadcasting systems will be essential parts of 4G, as depicted in Figure 1.8.

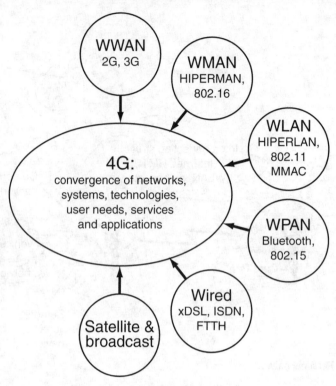

Figure 1.8 An illustration of the 4G concept. (HIPERMAN: high performance metropolitan area network; HIPERLAN: high performance local area network; MMAC: multimedia mobile access communication; xDLS: x-digital subscriber line; ISDN: Integrated Service Digital Network; FTTH: fiber to the home.)

1.6 Review of the Book

This book shows the present and future trends that WPAN infrastructures will follow with regard to synergies with 4G systems. It is organized into three main parts: the first one is devoted to the state of the art of current and near future WPANs, the second part presents future trends and research topics, and part three presents the authors' vision of the future WPANs and 4G.

Part 1 consists of Chapters 2 through 4. Chapter 2 presents an overview of WLANs. WLANs provide a new forum of access technology in the LAN world. The new access technology fulfills several practical requirements (increased mobility and flexibility), but several technical problems still remain unsolved. The problems of WLANs are discussed here. Chapter 3 presents a technical overview of Bluetooth. The goal of Bluetooth is to enable users to connect a wide range of computing and telecommunication devices easy and simply, without the need to buy, carry, and connect cables. Chapter 4 introduces IEEE 802.15, which is a collection of new WPAN standards, ranging from the classical Bluetooth to (yet to be finalized) very high data rate WPANs (up to 1 Gbps).

Part 2 consists of Chapters 5 through 8. Chapter 5 introduces the main motivations behind the PAN and PN concepts, introducing the main challenges and issues. In particular, it introduces the pervasive computing concept, a paradigm where computation power, as well as sensors, displays, and so forth, are available everywhere, embedded in the environment and easily accessible. Chapter 6 presents, in relation to the challenges identified in Chapter 5, a small subset of the most relevant research efforts that are taking place in the PAN and pervasive computing (or ubiquitous computing) domain. Chapter 7 introduces mobile ad hoc networks (MANET), an essential component of WPANs, as well as multiple antenna technologies that can enhance and help to solve MANET's capacity problems. The basic concept of security and its challenges are discussed in Chapter 8. Security in heterogeneous networks is one of the biggest technology challenges of the 21st century. Like for many worrying fields, however, everybody talks about it, but the ones who should feel the most affected do not seem to have detected the scale of the potential disaster.

Chapters 9 and 10 make up Part 3. There is an unpredictable future for wireless communication in general. However, the authors feel that the WPAN will play a very important role in the development of 4G. This idea is introduced in Chapter 9. Chapter 10 introduces applications of personal networks.

References

[1] IEEE 802.15, http://grouper.ieee.org/groups/802/15.

[2] Howitt, I., "WLAN and WPAN Coexistence in UL Band," *IEE Trans. on Vehicular Technology*, Vol. 50, No. 4, July 2001.

[3] Lansford, J., A. Stephens, and R. Nevo, "Wi-Fi (802.11b) and Bluetooth: Enabling Coexistence," *IEEE Network Magazine*, Vol. 15, No.5, September/October 2001, pp. 20–27.

[4] Prasad, R., "Basic Concept of Personal Area Networks," *WWRF, Kick-off Meeting*, Munich, Germany, 2001.

[5] Niemegeers, I. G., R. Prasad, and C. Bryce, "Personal Area Networks," *WWRF Second Meeting*, Helsinki, Finland, May 10–11, 2001.

[6] Prasad, R., "60 GHz Systems and Applications," *Second Annual Workshop on 60 GHz WLAN Systems and Technologies*, Kungsbacka, Sweden, May 15–16, 2001.

[7] Prasad, R., and L. Gavrilovska, "Personal Area Networks," *Proceedings EUROCON*, Bratislava, Slovakia, Vol. 1, July 2001, pp. III–VIII.

[8] Prasad, R., and L. Gavrilovska, "Research Challenges for Wireless Personal Area Networks," *Proceedings of Third International Conference on Information, Communications and Signal Processing (ICICS)*, Singapore, October 2001.

[9]   Zimmerman, T. G., "Personal Area Networks (PAN): Near-Field Intra-Body Communication," M.S. Thesis, MIT Media Lab., Cambridge, Massachusetts, 1995.

[10]   IBM PAN, http://www.almaden.ibm.com/cs/user/pan/pan.html.

[11]   MIT Oxygen project, http://oxygen.lcs.mit.edu/.

[12]   PicoRadio, http://www.gigascale.org/picoradio/.

[13]   IrDA Standards, http://www.irda.com/.

[14]   Bluetooth, http://www.bluetooth.com/.

[15]   "Advances in Mobile Ad Hoc Networking," special issue of *IEEE Personal Communications*, Vol. 8, No. 1, 2001.

[16]   Foerster, J., et al., "Ultra-Wideband Technology for Short- or Medium-Range Wireless Communications," *Intel Technology Journal*, Q2, 2001, http://intel.com/technology/itj/q22001/articles/art_4.htm.

[17]   Niemegeers, I. G., and S. M. Heemstra de Groot, "From Personal Area Networks to Personal Networks: A User Oriented Approach," *Wireless Personal Communications*, Vol. 22, August 2002, pp. 175–186.

[18]   Mohr, W., et al., Editors, "The Book of Visions 2001," Version 1.0, Wireless Strategic Initiative, December 2001.

Part 1
Tales of Ancient Times

2

WLANs

2.1 Introduction

The past decade has witnessed major changes in the types of communication services provided to users as well as in the infrastructure used to support them. Besides present-day telephony, Internet access, applications with remote servers, video on demand, and interactive multimedia are just a few examples of such services. Internet access is the service that has captured the biggest market and which enjoys maximum penetration; this is shown in Figure 2.1. Wireline communications networks that provide these services are commonly known as wide area networks (WANs) and local area networks (LANs).

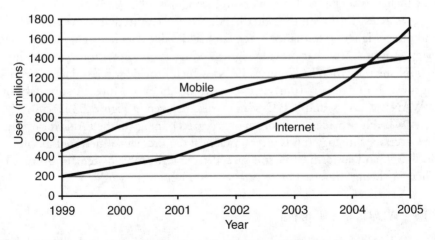

Figure 2.1 Growth in wireless and Internet. (*From:* Ericsson, "Emerging Markets in Telecommunications," http://connect.com, May 2001.)

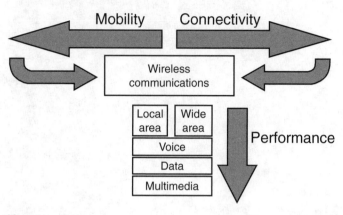

Figure 2.2 Market trend.

The overall market demand is basically connectivity, mobility, and performance. Wireline services can provide connectivity and performance but not mobility together with connectivity; this market demand is depicted in Figure 2.2. Wireless communications is the solution to these requirements. As a result, along with the growth of the Internet, there has been tremendous growth in the field of wireless communications (Figure 2.1). This has also been due to other inherent benefits of wireless, namely decreased wiring complexity, increased flexibility, and ease of installation. The main reason behind the growth of wireless has been wireless WANs (WWANs) or mobile technologies based on second generation (2G / 2.5G) standards like Global System for Mobile Communications (GSM) and Personal Digital Cellular (PDC). These technologies mainly provide voice services and some data services at low data rates. 3G systems provide higher data rates with a maximum throughput of 2 Mbps.

Wireless LANs, on the other hand, provide connectivity, lower mobility, and much higher performance in terms of achievable data rate. They are mainly extensions of LANs that provide high-speed data services with lower mobility. Complementary to WLANs are WPANs, which provide wireless data networking in a short range (~10m) at data rates of about 1 Mbps. A summary of WWANs, WLANs, and WPANs standards are given in Figure 2.3 [1–58].

WLANs provide a new forum of access technology in the LAN world. The new access technology fulfils several practical requirements (increased mobility, flexibility, and so forth), but still several technical problems remain unsolved. These are being addressed by researchers throughout the world.

2.1.1 WLANs in a Nutshell

Most WLANs operate using either radio or infrared techniques. Each approach has its own attributes that satisfy different connectivity requirements. The majority

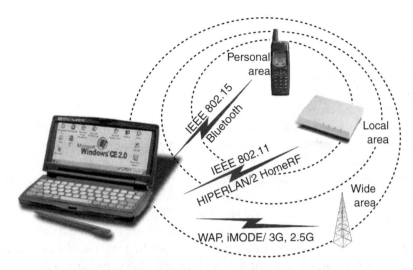

Figure 2.3 Wide area, local area, and personal area wireless technologies.

of these devices are capable of transmitting information up to several 100 meters in an open environment. Figure 2.4 gives an example of a WLAN interfacing with a wired network. The components of WLANs consist of a wireless network interface card, known as a station (STA), and a wireless bridge, referred to as an access point (AP). The AP interfaces the wireless network with the wired network (e.g., Ethernet LAN) [10–16, 28].

The most widely used WLANs use radio waves at the frequency band of 2.4 GHz, also known as the industrial, scientific, and medical (ISM) band. The worldwide availability of the ISM bands, shown in Figure 2.5 and in Appendix 2A: ISM Bands, has made unlicensed spectrum available and promoted

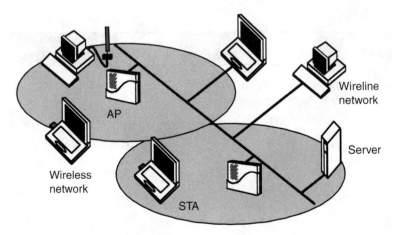

Figure 2.4 A wireless local area network.

significant interest in the design of WLANs. An advantage of radio waves is that they can provide connectivity for non-line-of-sight situations. A disadvantage of radio waves is that the electromagnetic propagation may cause interference with equipment working at the same frequency. Because radio waves propagate through walls, security might also be a problem. Further details of ISM band standards are given in Appendix 2B.

WLANs based on radio waves usually use spread spectrum technology. Spread spectrum *spreads* the signal power over a wide band of frequencies, which makes the data much less susceptible to electrical noise than conventional radio modulation techniques. Spread spectrum modulators use one of two methods to spread the signal over a wider spectrum: frequency hopping spread spectrum (FHSS) or direct sequence spread spectrum (DSSS) [29].

FHSS works very much as the name implies. It takes the data signal and modulates it with a carrier signal that hops from frequency to frequency as a function of time over a wide band of frequencies. DSSS combines a data signal at a sender with a higher data rate bit sequence, thus spreading the signal over the whole frequency band [28, 29]. Infrared LANs provide an alternative to radio wave-based WLANs. Although infrared has its benefits, it is not suitable for many mobile applications due to its line-of-sight requirement [28].

The first WLAN products appeared in the market around 1990, although the concept of WLANs was known for some years. The next generation of WLAN products was implemented on PCMCIA cards (also called PC card) that are used in laptop computers and portable devices. In recent years several WLAN standards have come into being. IEEE 802.11-based WLAN [10, 11, 28] was the first and most prominent in the field. IEEE 802.11 has different

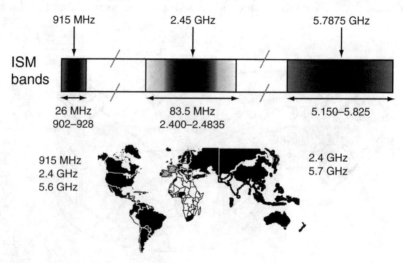

Figure 2.5 Worldwide availability of ISM bands.

physical layers working in 2.4 and 5 GHz. In 1999 the Wireless Ethernet Compatibility Alliance (WECA) was started. The purpose of WECA was to bring interoperability amongst IEEE 802.11 products of various vendors. The alliance developed a wireless-fidelity (Wi-Fi) interoperability test and provided logos for products that had passed the test. Today Wi-Fi has become a synonym of IEEE 802.11 and the alliance is now named the Wi-Fi Alliance.

Other WLAN standards are HomeRF [13] (now considered to be dead), which was dedicated to the home market based on FHSS, and high performance LAN (HIPERLAN) Type 2 [12, 17, 21, 28], which works in the 5-GHz band using orthogonal frequency division multiplexing (OFDM).

The exponential growth of the Internet and wireless has brought about tremendous changes in LAN technology. WLAN technology is becoming more and more important. Although WLAN came into being since the beginning of 1990s, the market has just started opening and the technology is still ripening. WLANs are to be used in several environments like home, office, and public hot spots, to name a few. This is depicted in Figure 2.6.

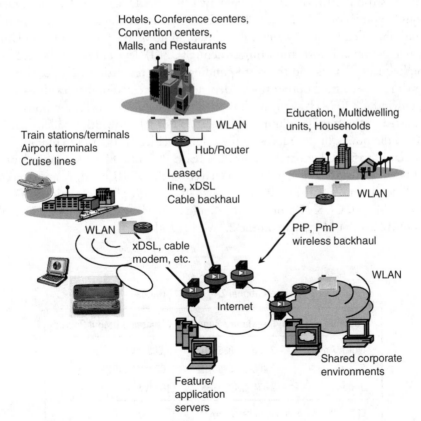

Figure 2.6 Usage environments for WLANs.

2.1.2   IEEE 802.11, HIPERLAN/2, and MMAC Wireless LAN Standards

Since the early 1990s, WLAN for the 900-MHz, 2.4-GHz, and 5-GHz ISM bands have been available, based on a range of proprietary techniques. In June 1997, the Institute of Electrical and Electronics Engineers (IEEE) approved an international interoperability standard, IEEE 802.11 [10]. The standard specifies both medium access control (MAC) procedures and three different physical layers (PHY). There are two radio-based PHYs using the 2.4-GHz band. The third PHY uses infrared light. All PHYs support a data rate of 1 Mbps and optionally 2 Mbps. The 2.4-GHz frequency band is available for license-exempt use in Europe, the United States, and Japan. Table 2.1 lists the available frequency band and the restrictions to devices that use this band for communications.

In July 1998, the PHY extension was a 2.4-GHz extension that employed complementary code keying (CCK), which is sometimes called HS/DSSS or IEEE 802.11b. CCK abandons the fixed, 11-bit Barker spreading of DSSS in favor of spreading codes that are related to the symbols they are spreading. This PHY can work at 11 and 5.5 Mbps with a fall back to 1- and 2-Mbps DSSS standard based on IEEE 802.11. In January 1997, the U.S. Federal Communications Commission (FCC) made 300 MHz of spectrum available in the 5.2-GHz band, intended for use by a new category of unlicensed equipment called Unlicensed National Information Infrastructure (UNII) devices [54]. Table 2.2 lists the frequency bands and the corresponding power restrictions. Notice that the maximum permitted output power depends on the emission bandwidth—for a bandwidth of 20 MHz, you are allowed to transmit at the maximum power levels listed in the middle column of Table 2.2. For a bandwidth smaller than 20 MHz, the power limit reduces to the value specified in the right column. The standard for 5.2-GHz band is known is IEEE 802.11a. 802.11a employs only OFDM (up to 54 Mbps) as the PHY layer on the UNII band. A more recent PHY development is IEEE 802.11g, which uses all three PHY techniques: DSSS (1 or 2 Mbps), CCK (5.5 or 11 Mbps), and a down-banded form of 802.11a's OFDM (6 to 54 Mbps); all on the 2.4-GHz band.

Table 2.1
International 2.4-GHz ISM Bands

Location	Regulatory Range	Maximum Output Power
North America	2.400–2.4835 GHz	1,000 mW
Europe	2.400–2.4835 GHz	100 mW (EIRP)
Japan	2.471–2.497 GHz	10 mW

EIRP: effective isotropic radiated power.

Table 2.2
United States 5.2-GHz UNII Band

Location	Maximum Output Power	Minimum of
5.150–5.250 GHz	50 mW	4 dBm + $10\log_{10}B$*
5.250–5.350 GHz	250 mW	11 dBm + $10\log_{10} B$
5.725–5.825 GHz	1,000 mW	17 dBm + $10\log_{10} B$

B is the −26-dB emission bandwidth in megahertz.

Like the IEEE 802.11 standard, the European ETSI HIPERLAN type 1 standard [55] specifies both MAC and PHY. Unlike IEEE 802.11, however, no HIPERLAN type 1 compliant product is available in the marketplace. A newly formed ETSI working group called Broadband Radio Access Networks (BRAN) is now working on extensions to the HIPERLAN standard. Three extensions are under development: HIPERLAN/2, a wireless indoor LAN with a QoS provision; HiperLink, a wireless indoor backbone; and HiperAccess, an outdoor, fixed wireless network providing access to a wired infrastructure.

In Japan, equipment manufacturers, service providers, and the Ministry of Post and Telecommunications are cooperating on the Multimedia Mobile Access Communication (MMAC) project to define new wireless standards similar to those of IEEE 802.11 and ETSI BRAN. Additionally, MMAC is also looking into the possibility for ultra-high-speed wireless indoor LANs supporting large-volume data transmission at speeds up to 156 Mbps using frequencies in the 30- to 300-GHz band.

In July 1998, the IEEE 802.11 standardization group decided to select OFDM as the basis for their new 5-GHz standard, targeting a range of data rates from 6 to 54 Mbps [56, 57]. This new standard is the first one to use OFDM in packet-based communications; the use of OFDM until now was limited to continuous transmission systems like digital audio broadcasting (DAB) and digital video broadcasting (DVB). Following the IEEE 802.11 decision, European Telecommunications Standards Institute (ETSI) BRAN and MMAC also adopted OFDM for their physical layer standards. The three bodies have worked in close cooperation since then to make sure that differences among the various standards are kept to a minimum, thereby enabling the manufacturing of equipment that can be used worldwide.

The focus of this section is on the physical layer side. In the case of the IEEE 802.11 standard, the MAC layer for the higher data rates remains the same as for the currently supported 1- and 2-Mbps rates. A description of this MAC can be found in [57].

2.1.2.1   OFDM Parameters

Table 2.3 lists the main parameters of the draft OFDM standard. A key parameter that largely determined the choice of the other parameters is the guard interval of 800 ns. This guard interval provides robustness to root mean square (rms) delay spreads up to several hundreds of nanoseconds, depending on the coding rate and modulation used. In practice, this means that the modulation is robust enough to be used in any indoor environment, including large factory buildings. It can also be used in outdoor environments, although directional antennas may be needed in this case to reduce the delay spread to an acceptable amount and increase the range.

To limit the relative amount of power and time spent on the guard time to 1 dB, the symbol duration chosen is 4 μs. This also determines the subcarrier spacing at 312.5 kHz, which is the inverse of the symbol duration minus the guard time. By using 48 data subcarriers, uncoded data rates of 12 to 72 Mbps can be achieved by using variable modulation types from BPSK to 64-QAM. In addition to the 48 data subcarriers, each OFDM symbol contains an additional four pilot subcarriers, which can be used to track the residual carrier frequency offset that remains after an initial frequency correction during the training phase of the packet.

To correct for subcarriers in deep fades, forward error correction (FEC) across the subcarriers is used with variable coding rates, giving coded data rates from 6 to 54 Mbps. Convolutional coding is used with the industry standard rate 1/2, constraint length 7 code with generator polynomials (133,171). Higher coding rates of 2/3 and 3/4 are obtained by puncturing the rate 1/2 code. The 2/3-rate is used together with 64-QAM only to obtain a data rate of 48 Mbps. The 1/2-rate is used with BPSK, QPSK, and 16-QAM to give rates of 6, 12, and

Table 2.3
Main Parameters of the OFDM Standard

Data Rate	6, 9, 12, 18, 24, 36, 48, 54 Mbps
Modulation	BPSK, QPSK, 16-QAM, 64-QAM
Coding Rate	1/2, 2/3, 3/4
Number of Subcarriers	52
Number of Pilots	4
OFDM Symbol Duration	4 μs
Guard Interval	800 ns
Subcarrier Spacing	312.5 kHz
−3-dB Bandwidth	16.56 MHz
Channel Spacing	20 MHz

24 Mbps, respectively. Finally, the 3/4-rate is used with BPSK, QPSK, 16-QAM, and 64-QAM to give rates of 9, 18, 36, and 54 Mbps, respectively.

2.1.2.2 Differences Between IEEE 802.11, HIPERLAN/2, and MMAC

The main differences between IEEE 802.11 and HIPERLAN/2 (which is standardized by ETSI BRAN [58]) are in the MAC. IEEE 802.11 uses a distributed MAC based on carrier sense multiple access with collision avoidance (CSMA/CA), while HIPERLAN/2 uses a centralized and scheduled MAC, based on wireless asynchronous transfer mode (ATM). MMAC supports both of these MACs. As far as the physical layer is concerned, there are a few relatively minor differences between IEEE 802.11a and HIPERLAN/2, which are summarized below. Note that unlike IEEE 802.11, HIPERLAN/2 has only one PHY using OFDM in the 5.2-GHz band.

HIPERLAN uses different training sequences. The long training symbol is the same as for IEEE 802.11, but the preceding sequence of short training symbols is different. A downlink transmission starts with 10 short symbols as IEEE 802.11, but the first 5 symbols are different in order to detect the start of the downlink frame. The rest of the packets in the downlink frame do not use short symbols, only the long training symbol. Uplink packets may use 5 or 10 identical short symbols, with the last short symbol being inverted.

HIPERLAN uses extra puncturing to accommodate the tail bits to keep an integer number of OFDM symbols in 54-byte packets. This extra puncturing operation punctures 12 bits out of the first 156 bits of a packet.

In the case of 16-QAM, HIPERLAN uses a coding rate of 9/16 instead of 1/2—giving a bit rate of 27 instead of 24 Mbps—to get an integer number of OFDM symbols for packets of 54 bytes. The rate 9/16 is made by puncturing 2 out of every 18 encoded bits.

Both IEEE 802.11 and HIPERLAN scramble the input data with a length 127 pseudo random sequence, but the initialization is different. IEEE 802.11 initializes with 7 random bits, which are inserted as the first 7 bits of each packet. In HIPERLAN, the scrambler is initialized by {1, 1, 1} plus the first 4 bits of the broadcast channel at the beginning of a MAC frame. The initialization is identical for all packets in a MAC frame.

HIPERLAN devices have to support power control in the range of −15 to 30 dBm with a step size of 3 dB.

Dynamic frequency selection is mandatory in Europe over a range of at least 330 MHz for indoor products and 255 MHz (upper band only) for outdoor products. This means that indoor products have to support a frequency range from 5.15 to at least 5.6 GHz, covering the entire lower band and a part of the European upper band. Dynamic frequency selection was included to avoid the need for frequency planning and to provide coexistence with radar systems that operate in the upper part of the European 5-GHz band.

2.2 MAC in WLAN Standards

The MAC protocols form the basis of efficient use of a channel, be it wireline or wireless. When numerous users desire to transmit in a channel at the same time, conflicts occur, so there must be procedures on how the available channel capacity is allocated. These procedures constitute the MAC protocol rules each user has to follow in accessing the common channel [30]. The channel thus becomes a shared resource whose allocation is critical for proper functioning of the network. With the boom of WLANs, an efficient MAC has become a must.

Before designing an appropriate MAC protocol, one has to understand the wireless network under discussion [30–32]. The first thing that should be understood is the duplexing scheme used by a system and also the network architecture. A MAC protocol is dependent on these two issues.

Duplexing refers to mechanisms for wireless devices to send and receive. There are two duplexing methods, time based or frequency based. Sending and receiving data in the same frequency at different time periods is known as time division duplex (TDD), while sending and receiving data in same time and different frequency is known as frequency division duplex (FDD).

A wireless network can be distributed or centralized. Distributed networks are those where each device accesses the medium individually and transmits the data without any central control. Distributed network architecture requires the same frequency and thus makes use of TDD. IEEE 802.11 is an example of distributed network architecture. Centralized network architecture has one network element, which controls the communication of various devices. Such network architecture can make use of both TDD and FDD. HIPERLAN/2 is an example of centralized network architecture.

In the following sections the MAC protocols in IEEE 802.11 [33, 34] and HIPERLAN/2 [35, 36] are discussed. IEEE 802.11 is the most commonly used WLAN, and it is explained in more detail.

2.2.1   IEEE 802.11

Standardization of IEEE 802.11 was done to satisfy the needs of wireless data networking. CSMA/CA was the MAC protocol adopted by IEEE 802.11 [3, 10]. Wherein, the basic channel access method is random back-off CSMA with a MAC-level acknowledgment. A CSMA protocol requires the STA to listen before talk. In this protocol only one user can access the medium at a time while the system is mostly used for low data rate applications (Internet access, e-mail, and so forth). The IEEE 802.11 basic medium access behavior allows interoperability between compatible PHYs through the use of the CSMA/CA protocol and a random back-off time following a busy medium condition. In addition, all traffic uses immediate positive acknowledgment (ACK frame), where the sender schedules a retransmission if no ACK is received. The IEEE 802.11 CSMA/CA protocol is designed to reduce the collision probability between multiple stations accessing the medium at the point in time where collisions would most likely occur. Collisions are most likely to happen just after the medium becomes free (i.e., just after busy medium conditions). This is because multiple stations would have been waiting for the medium to become available again. Therefore, a random back-off arrangement is used to resolve medium contention conflicts. The IEEE 802.11 MAC also describes the way beacon frames are sent by the AP at regular intervals (like 100 ms) to enable stations to monitor the presence of the AP. The MAC also gives a set of management frames that allow a station to actively scan for other APs on any available channel. Based on this information the station may decide on the best-suited AP. In addition, the 802.11 MAC defines special functional behavior for the fragmentation of packets, medium reservation via request-to-send/clear-to-send (RTS/CTS) polling interaction and point coordination (for time-bounded services) [33].

The MAC sublayer is responsible for the channel allocation procedures, protocol data unit (PDU) addressing, frame formatting, error checking, and fragmentation and reassembly. The transmission medium can operate in the contention mode exclusively, requiring all stations to contend for access to the channel for each packet transmitted. The medium can also alternate between the contention mode, known as the contention period (CP), and a contention-free period (CFP). During the CFP, medium usage is controlled (or mediated) by the AP, thereby eliminating the need for stations to contend for channel access. IEEE 802.11 supports three different types of frames: management, control, and data. The management frames are used for station association and disassociation with the AP, timing and synchronization, and authentication and de-authentication. Control frames are used for handshaking during the CP, for positive acknowledgments during the CP, and to end the CFP. Data frames are used for the transmission of data during the CP and CFP; these can be combined with polling and acknowledgments during the CFP.

Since the contention-free mode is not used, this section will discuss the contention mode of IEEE 802.11 MAC, which is also known as the distributed coordination function (DCF). The RTS/CTS mechanism of IEEE 802.11 is not discussed in this chapter. The IEEE 802.11 MAC discussed here is the original MAC and not IEEE 802.11e and i, where work on QoS and security, respectively, are presented.

2.2.1.1 Distributed Coordination Function

The DCF is the fundamental access method used to support asynchronous data transfer on a best effort basis. As identified in the IEEE 802.11 specification [3, 10], all stations must support the DCF. The DCF operates solely in the ad hoc network, and it either operates solely or coexists with the PCF in an infrastructure

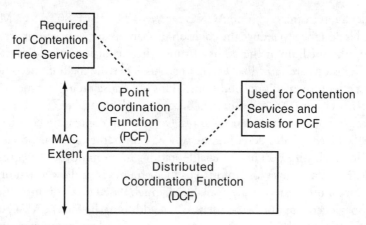

Figure 2.7 MAC architecture.

network. The MAC architecture is depicted in Figure 2.7, where it is shown that the DCF sits directly on top of the physical layer and supports contention services. Contention services imply that each station with a packet queued for transmission must contend for access to the channel for each frame transmitted. Contention services promote fair access to the channel for all stations [33].

The DCF is based on CSMA/CA. In IEEE 802.11, carrier sensing is performed at both the air interface, referred to as physical carrier sensing, and at the MAC sublayer, referred to as virtual carrier sensing. Physical carrier sensing detects the presence of other IEEE 802.11 WLAN users by analyzing all detected packets, and it also detects activity in the channel via relative signal strength from other sources.

A source station performs virtual carrier sensing by sending packet duration information in the header of RTS, CTS, and data frames. A packet is a complete data unit that is passed from the MAC sublayer to the physical layer. The packet contains header information, payload, and a 32-bit cyclic redundancy check (CRC). The duration field indicates the amount of time (in microseconds) after the end of the present frame the channel will be utilized to complete the successful transmission of the data or management frame. Stations in the basic service set (BSS) use the information in the duration field to adjust their network allocation vector (NAV), which indicates the amount of time that must elapse until the current transmission session is complete and the channel can be sampled again for idle status. The channel is marked busy if either the physical or virtual carrier sensing mechanisms indicate the channel is busy.

Priority access to the wireless medium is controlled through the use of interframe space (IFS) time intervals between the transmission of frames. The IFS

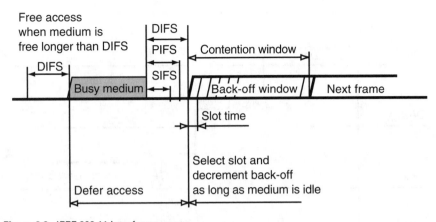

Figure 2.8 IEEE 802.11 interframe space.

intervals are mandatory periods of idle time on the transmission medium. Three IFS intervals (see Figure 2.8) are specified in the standard: short IFS (SIFS), point coordination function IFS (PIFS), and DCF-IFS (DIFS). The SIFS interval is the smallest IFS, followed by PIFS and DIFS, respectively. Stations only required to wait a SIFS period have priority access over those stations required to wait a PIFS or DIFS period before transmitting; therefore, SIFS has the highest priority access to the communications medium. For the basic access method, when a station senses the channel is idle, the station waits for a DIFS period and samples the channel again. If the channel is still idle, the station transmits a MAC layer protocol data unit (MPDU). The receiving station calculates the checksum and determines whether the packet was received correctly. Upon receipt of a correct packet, the receiving station waits a SIFS interval and transmits a positive ACK frame back to the source station, indicating that the transmission was successful. Figure 2.9 is a timing diagram illustrating the successful transmission of a data frame. When the data frame is transmitted, the duration field of the frame is used to let all stations in the BSS know how long the medium will be busy. All stations hearing the data frame adjust their NAV based on the duration field value, which includes the SIFS interval and the ACK following the data frame.

The collision avoidance portion of CSMA/CA is performed through a random back-off procedure. If a station with a frame to transmit initially senses the channel to be busy, the station waits until the channel becomes idle for a DIFS period, and then computes a random back-off time. For the IEEE 802.11, time is slotted in time periods that correspond to a Slot_Time. The Slot_Time used in IEEE 802.11 is much smaller than an MPDU and is used to define the IFS intervals and determine the back-off time for stations in the CP. The Slot_Time

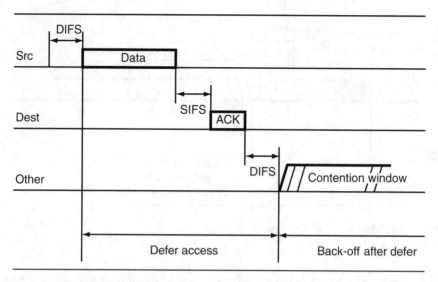

Figure 2.9 Transmission of a MPDU without RTS/CTS.

is different for each physical layer implementation. The random back-off time is an integer value that corresponds to a number of time slots. Initially, the station computes a back-off time in the range 0 to 7. After the medium becomes idle after a DIFS period, stations decrement their back-off timer until the medium becomes busy again or the timer reaches zero. If the timer has not reached zero and the medium becomes busy, the station freezes its timer. When the timer is finally decremented to zero, the station transmits its frame. If two or more stations decrement to zero at the same time, a collision will occur which leads to missing ACKs, and each station will have to generate a new back-off time in the range 0 to 63 (for 802.11b and 0 to 31 for 802.11a) times the Slot_Time period. The generated back-off time corresponds to a uniform distributed integer multiple of Slot_Time periods. For the next retransmission attempt, the back-off time grows to the range 0 to 127 (for 802.11b, and 0-63 for 802.11a) Slot_Time periods and so on with a maximum in the range of 0 to 1,023. The idle period after a DIFS period is referred to as the contention window (CW). The advantage of this channel access method is that it promotes fairness among stations, but its weakness is that it probably could not support time-bound services. Fairness is maintained because each station must recontend for the channel after every transmission of an MPDU. All stations have equal probability of gaining access to the channel after each DIFS interval. Time-bounded services typically support applications such as packetized voice or video that must be maintained with a specified minimum delay. With DCF, there is no mechanism to guarantee minimum delay to stations supporting time-bounded services.

2.2.2 HIPERLAN/2

The MAC in HIPERLAN/2 is a part of the data link control (DLC) layer together with other functions like error control (EC). A brief description of the MAC layer and frames of HIPERLAN/2 are given in this section [35, 36].

2.2.2.1 MAC Layer

The MAC scheme of HIPERLAN/2 is based on a central controller, which is located at the AP. The core task of the central controller is to determine the direction of information flow between the controller and the terminal at any point of time. A MAC frame consists of control and data blocks. The central controller decides which terminal or group of terminals is allowed to transmit in a slot of the frame. The medium access scheme is classified as load adaptive time division multiple access (TDMA). Each user shall be assigned zero, one, or several slots in a frame. In general, the number of slots assigned to an individual user varies from frame to frame and depends on the actual bandwidth request of the terminal. The uplink and downlink packets are sent on the same frequency channel in a TDD mode.

Random access slots are provided to allow STAs to get associated with the controller. In this bootstrap phase, data is transmitted in a contention-based mode. Since collisions may occur, a collision resolution algorithm is applied.

Uplink signaling of resource describes the state of the input queues of a STA to the central controller. The AP collects these requests from all associated STAs and uses this data to schedule the uplink access times. The results of the scheduling process are signaled via the frame control channel (i.e., a description of the exact frame structure and slot allocation is contained in the frame control channel). These control data are valid for the ongoing frame. Additional tasks include multiplexing and demultiplexing of logical channels, service requesting and service granting, and means of medium access control.

2.2.2.2 MAC Frames

The MAC frame structure (Figure 2.10) comprises time slots for broadcast control (BCH), frame control (FCH), access feedback control (ACH), and data transmission in downlink (DL), uplink (UL), and directlink (DiL) phases, which are allocated dynamically depending on the need for transmission resources. An STA first has to request capacity from the AP in order to send its data. This can be done in the random access channel (RCH), where contention for the same time slot is allowed.

DL, UL, and DiL phases consist of two types of PDUs: long PDUs and short PDUs. The long PDUs (Figure 2.11) have a size of 54 bytes and contain control or user data. The payload is 49.5 bytes, and the remaining 4.5 bytes are used for the PDU type (2 bits), a sequence number (10 bits, SN), and CRC (24

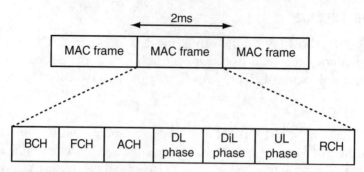

Figure 2.10 The HIPERLAN/2 MAC frame.

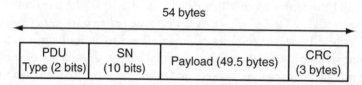

Figure 2.11 Format of the long PDUs.

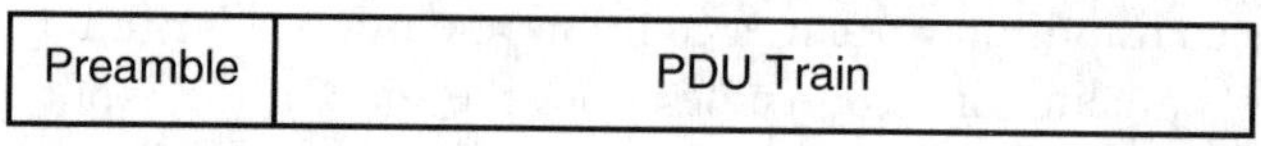

Preamble	PDU Train

Figure 2.12 Format of PDU train.

bits). Long PDUs are referred to as the long transport channel (LCH). Short PDUs contain only control data and have a size of 9 bytes. They may contain resource requests or ARQ messages, and they are referred to as the short transport channel (SCH).

Traffic from multiple connections to and from one STA can be multiplexed onto one PDU train, which contains long and short PDUs. A physical burst is composed of the PDU train payload and a preamble; and this is the unit to be transmitted via the physical layer (Figure 2.12).

2.3 QoS Over Wireless LANS

Quality of service is becoming an increasingly important element of any communications system. In the simplest sense, QoS means providing a consistent, predictable data delivery service, in other words, satisfying the customer application requirements. Providing QoS means providing real-time (e.g., voice) as well as nonreal-time services. Voice communication is the primary form of service required by human users. The public telephone network and the equipment that makes it possible are taken for granted in most parts of the world. Availability of a telephone and access to a low-cost, high-quality worldwide network is considered to be essential in this modern society (telephones are expected to work even when the power is off).

Support for voice communications using the Internet Protocol (IP), which is usually just called voice over IP (VoIP), has become especially attractive given the low cost, flat rate pricing of the public Internet. VoIP can be defined as the ability to make telephone calls [i.e., to do everything that can be done today with the Public Switched Telecommunications/Telephone Network (PSTN)] over the IP-based data networks with a suitable QoS. This is desirable because of its much superior cost benefit ratio compared to the PSTN. Equipment producers see VoIP as a new opportunity to innovate and compete. The challenge for them is turning this vision into reality by quickly developing new VoIP-enabled equipment that is capable of providing toll quality service. For Internet service providers (ISPs) the possibility of introducing usage-based pricing and increasing their traffic volumes is very attractive. Both the ISPs and the network manufacturers face the challenge of developing and producing solutions that can provide the required voice quality. Users are seeking new types of integrated voice/data applications as well as cost benefits.

Since WLANs are extensions of the IP to the wireless realm, it is necessary to have a voice over WLAN (VoWLAN) protocol that will fulfill this requirement. A complete system for voice over WLAN, IP to Plain Old Telephone System (POTS), is depicted in Figure 2.13. Successfully delivering VoWLAN presents a tremendous opportunity; however, implementing the products is not as straightforward a task as it may first appear.

The following sections presents the QoS mechanisms adopted in the present IEEE 802.11e draft standard (which is the IEEE 802.11 standard for MAC enhancements for QoS).

2.3.1   IEEE 802.11e

IEEE 802.11e provides MAC enhancements to support LAN applications with QoS requirements. The QoS enhancements are available to the QoS enhanced stations (QSTAs) associated with a QoS enhanced access point (QAP) in a QoS enabled network. A subset of the QoS enhancements may be available for use between QSTAs. A QSTA may associate with a non-QoS AP in a non-QoS network [6, 7]. Non-QoS STAs may associate with a QAP. Figure 2.14 shows the MAC architecture of the IEEE 802.11e.

The enhancements that distinguish the QSTAs from non-QoS STAs and the QAPs from non-QoS APs comprise an integrated set of QoS-related formats and functions that are collectively termed the QoS facility. The quantity of

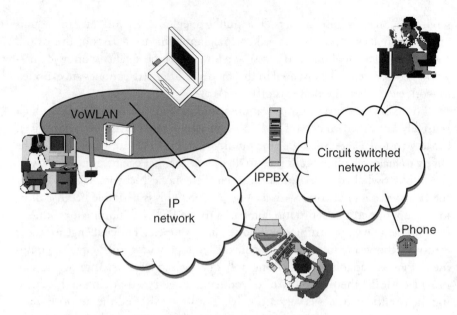

VoWLAN: Voice over Wireless Local Area Network
IP: Internet Protocol
PBX: Private Branch eXchange

Figure 2.13 Voice over WLAN, IP to POTS.

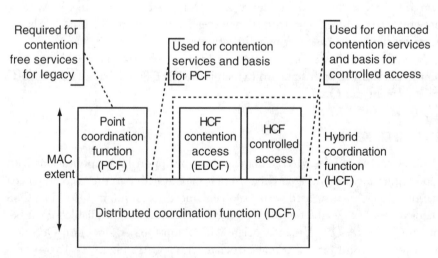

Figure 2.14 IEEE 802.11e MAC architecture.

certain, QoS-specific mechanisms may vary among QoS implementations, as well as between the QSTAs and the QAPs [7]. However, all service primitives, frame formats, coordination function and frame exchange rules, and management interface functions defined as part of the QoS facility are mandatory, with the exception of the group acknowledgement function defined [7], which is an option separate from the core QoS facilities and the presence of which is indicated by QSTAs separate from the core QoS facility.

The IEEE 802.11e standard provides two mechanisms for the support of applications with QoS requirements. The first mechanism, designated as the enhanced distributed coordination function (EDCF), is based on the differentiating priorities at which the traffic is to be delivered. This differentiation is achieved through varying the amount of time a station would sense the channel to be idle, the length of the contention window during back-off, or the duration for which a station may transmit once it has the channel access.

The second mechanism allows for the reservation of transmission opportunities with the hybrid coordinator (HC). A QSTA based on its requirements requests the HC for transmission opportunities—both for its own transmissions as well as transmissions from the HC to itself. The HC, based on an admission control policy, either accepts or rejects the request. If the request is accepted, it schedules transmission opportunities for the QSTA. For transmissions from the STA, the HC polls a QSTA based on the parameters supplied by the QSTA at the time of its request. For transmissions to the QSTA, the HC queues the frames and delivers them periodically, again based on the parameters supplied by the QSTA. This mechanism is expected to be used for applications such as voice and video, which may need a periodic service from the HC. This mechanism is a hybrid of several proposals studied by the standardization committee.

2.3.2 Interframe Spacing

The time interval between frames is called the IFS. A STA determines that the medium is idle through the use of the carrier sense function for the interval specified. Five different IFSs are defined to provide priority levels for access to the wireless media; they are listed in order, from the shortest to the longest—except for the arbitration interframe space (AISF). Figure 2.15 shows some of these relationships. The different IFSs are independent of the STA data rate:

- Short interframe space;
- PCF interframe space;
- DCF interframe space (DIFS);
- Arbitration interframe space (used by the QoS facility);
- Extended interframe space (EISF).

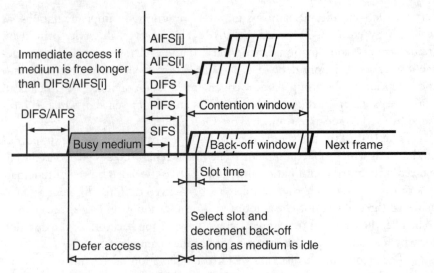

Figure 2.15 Interframe spacing for enhanced MAC.

The AIFS is to be used by QSTAs to transmit data and management frames. A QSTA using the EDCF is allowed a transmit opportunity (TxOP) for a particular traffic class (TC) if its carrier sense mechanism determines that the medium is idle at the TxAIFS(TC) slot boundary after a correctly received frame and the back-off time for that TC has expired.

2.3.3   Other QoS-Related Developments

IEEE 802.11e is also looking into admission control and scheduling; this will, of course, complete the picture for QoS support. In addition, the IEEE 802.11 standard is also looking into overlapping cell issues, which is very important for QoS especially for a system with very few and unlicensed nonoverlapping channels. IEEE 802.11f provides a recommendation for the interaccess point protocol (IAPP; approved as standard in July 2003) [15]—the transferring of context information from one AP to another will help in fast and seamless handover within an administrative domain in a IP subnet.

2.4   Security in IEEE 802.11

The use of WLANs is experiencing explosive growth and the once-imagined usage environments have already been reached: the realms of academia, enterprise, and public hotspots. Security is an issue, however, that can cause a major setback to the growth of WLANs [37–53]. This section discusses security issues addressed in the original IEEE 802.11 standard [38, 39, 46], as well as in IEEE 802.11i (standard for security enhancements) and the IEEE 802.11f recommendation (standard for fast mobility). (IEEE 802.11 uses the word *roaming*, which is the same as *handover* in mobile communications.)

2.4.1 Current IEEE 802.11 Resources

2.4.1.1 Authentication

IEEE 802.11 defines two subtypes of authentication service: open system and shared key [46]. Open system authentication is the simplest of the available authentication algorithms. Essentially it is a null authentication algorithm. Any STA that requests authentication with this algorithm may become authenticated if the recipient station is set to the open system authentication.

The shared key authentication supports authentication of the STAs as either a member of those who know a shared secret key or a member of those who do not. The IEEE 802.11 shared key authentication accomplishes this without the need to transmit the secret key in the clear; requiring the use of the wired equivalent privacy (WEP) mechanism. Therefore, this authentication scheme is only available if the WEP option is implemented. The required secret, shared key is presumed to have been delivered to participating STAs via a secure channel that is independent of IEEE 802.11. During the shared key authentication exchange, both the challenge and the encrypted challenge are transmitted. This facilitates unauthorized discovery of the pseudo random number (PRN) sequence for the key/initialization vector (IV) pair used for the exchange. Therefore the same key/IV pair for subsequent frames should not be used. The shared key authentication process is shown in Figure 2.16.

2.4.1.2 Wired Equivalent Privacy

The WEP algorithm is a form of electronic codebook in which a block of plaintext is bitwise XOR-ed with a pseudorandom key sequence of equal length. The key sequence is generated by the WEP algorithm.

Referring to Figure 2.17 and viewing from the left to right, the encipherment begins with a secret key that has been distributed to the cooperating STAs by an external key management service. WEP is a symmetric algorithm in which the same key is used for encipherment and decipherment.

The secret key is concatenated with an IV, and the resulting seed is an input to the pseudo random number generator (PRNG). The PRNG outputs a key sequence k of pseudorandom octets equal in length to the number of data octets that are to be transmitted in the MPDU plus 4 [since the key sequence is used to protect the integrity check value (ICV) as well as the data]. Two

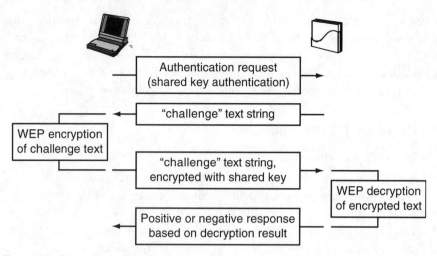

Figure 2.16 Shared key authentication.

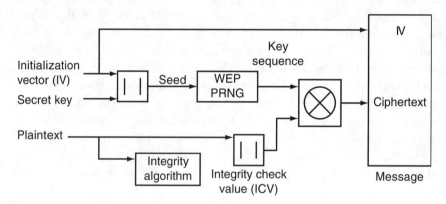

Figure 2.17 WEP encipherment block diagram.

processes are applied to the plaintext MPDU. To protect against unauthorized data modification, an integrity algorithm operates on the plaintext MPDU to produce an ICV. Encipherment is then accomplished by mathematically combining the key sequence with the plaintext concatenated with the ICV. The output of the process is a message containing the IV and ciphertext.

Referring to Figure 2.18 and viewing from the left to the right, the decipherment begins with the arrival of a message. The IV of the incoming message shall be used to generate the key sequence necessary to decipher the incoming message. Combining the ciphertext with the proper key sequence yields the original plaintext and the ICV. Correct decipherment shall be verified by performing the integrity check algorithm on the recovered plaintext and comparing the

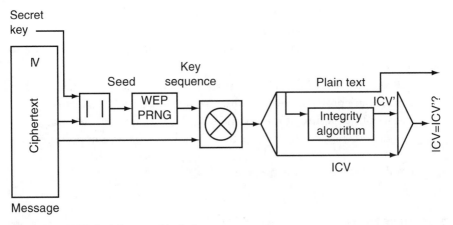

Figure 2.18 WEP decipherment block diagram.

output ICV′ to the ICV transmitted with the message. If ICV′ is not equal to ICV, the received MPDU is in error and an error indication is sent to the MAC management.

2.4.1.3 IEEE 802.11 Security Issues

The following security issues were known and identified as of this writing:

- *Shared key authentication:* Shared key authentication suffers a known-plaintext attack. Indeed, the plaintext (challenge) and cipher text (encrypted challenge) can be eavesdropped from the air. As xoring the plain-text and the cipher text provide the pseudo Random String, this string can be used in a new authentication even though the shared secret is not recovered.

- *Mutual authentication:* WEP provides no mutual authentication between the station and the access point. That is, the AP can authenticate the station, but not vice versa.

- *Key management:* There is no real key management in WEP, but two methods of using WEP keys are provided. The AP and the stations share the usage of the four (default) keys. The compromising of any of the nodes means a compromise of the wireless network. A key mappings table is used at the AP. In this method, each unique MAC address can have a separate key. The size of a key mappings table should be at least 10 entries according to the 802.11 specification; however, it is likely to be chip-set dependent. The use of a separate key for each user mitigates the known cryptographic attacks, but requires more effort on the manual key management. Since key distribution is not defined in WEP and can be done only manually, many of the organizations deploying wireless networks use either a permanent fixed cryptographic variable, or key, or no encryption at all.

- *Other problems:* Since WLAN APs are usually connected to the intranets, which are protected by firewalls, a compromise of WLAN can result in a serious exposure of the intranets. Using the same key for authentication and encryption increases the possibility of being compromised.

For details on security issues, refer to [47].

2.4.2   IEEE 802.11i

This section provides an overview of IEEE 802.11i (security enhancements of IEEE 802.11). A short description of IEEE 802.1X is also given as it plays an important role in future WLAN security solutions.

2.4.2.1   IEEE 802.11i

The IEEE 802.11i is a security enhancement standard published in July 2004. While the IEEE 802.11i standard are not available, the IEEE 802.11 vendors have been providing some security solutions to bridge the gap. These solutions include extending the WEP key size (which was adopted by the standard), providing RADIUS and MAC address based authentication, the IEEE 802.1X port based user authentication, and the AES based encryption.

Realizing the market situation, Wi-Fi (the IEEE 802.11 interoperability industry alliance) is introducing the Temporal Key Integrity Protocol (TKIP) as a simple but secure intermediary solution [11]. This solution is usually known as Wi-Fi protected access (WPA) and is already available from some vendors. The WPA provides enhanced data encryption through TKIP, user authentication via IEEE 802.1X and Extensible Authentication Protocol (EAP), and mutual authentication; for ease of transition, Wi-Fi certified products are software upgradeable.

The following gives an overview of the current IEEE 802.11i draft.

The Robust Security Network

IEEE 802.11i defines a robust security network (RSN). A RSN provides a number of additional security features not present in the basic IEEE 802.11 architecture. These features include:

- Enhanced authentication mechanisms for both the APs and the STAs;
- Key management algorithms;

- Dynamic, association-specific cryptographic keys;
- An enhanced data encapsulation mechanism.

An RSN makes extensive use of protocols above the IEEE 802.11 MAC layer to provide the authentication and the key management. This allows the IEEE 802.11 standard to take advantage of the work already done by other standardization bodies as well as to avoid duplicating functions at the MAC layer that are already performed at the higher layers. RSN introduces several new components into the IEEE 802.11 architecture:

- *IEEE 802.1X Port:* This is present on all the STAs in a RSN and resides above the 802.11 MAC. All data traffic that flows through the RSN MAC also passes through the IEEE 802.1X Port.
- *Authentication agent (AA):* This component resides on top of the IEEE 802.1X Port at each STA and provides for the authentication and the key management.
- *Authentication server (AS):* This is an entity that resides in the network and participates in the authentication of all STAs (including the APs). It may authenticate the elements of the RSN itself or it may provide material that the RSN elements can use to authenticate each other.

As IEEE 802.1X plays an important role in IEEE 802.11i, it is explained separately in a following section.

Security Goals

An RSN does not directly provide the services. Instead, an RSN uses the IEEE 802.1X to provide the access control and the key distribution, and confidentiality is provided as a side effect of the key distribution. Some of the security goals of IEEE 802.11i are as follows:

- *Authentication:* An RSN-capable IEEE 802.11 network also supports the upper layer authentication, based on IEEE 802.1X. The upper layer authentication utilizes the protocols above the MAC layer to authenticate the STAs and the network with one another.
- *Deauthentication:* In an RSN using the upper layer authentication, the deauthentication may result in the IEEE 802.1X controlled port for the station being disabled.
- *Privacy:* IEEE 802.11i provides three cryptographic algorithms to protect the data traffic. Two are based on the RC4 algorithm defined by Rivest, Shamir, and Adleman (RSA), and the third is based on the

Advanced Encryption Standard (AES). This standard refers to these as WEP, TKIP, and WRAP, respectively. A means is provided for the stations to select the algorithm to be used for a given association. The Wireless Robust Authenticated Protocol (WRAP), adopted by IEEE 802.11i, is based on the AES and the Offset Codebook (OCB).

- *Key distribution:* IEEE 802.11i supports two key distribution mechanisms. The first is the manual key distribution. The second is the automatic key distribution, which is available only in an RSN that uses IEEE 802.1X to provide the key distribution services. An RSN allows a number of authentication algorithms to be utilized. The standard does not specify a mandatory-to-implement upper layer authentication protocol.

- *Data origin authentication:* This mechanism is available only to stations using WRAP and TKIP. The data origin authenticity is only applicable to the unicast traffic.

- *Replay detection:* This mechanism is available only to stations using WRAP and TKIP.

IEEE 802.1X

IEEE 802.1X is the standard for port-based network access control, which applies to the IEEE 802.3 Ethernet, the Token Ring, and the WLAN [49]. Based on the Point-to-Point Protocol (PPP) EAP [50], IEEE 802.1X extends the EAP from PPP to the LAN applications. This standard defines the EAP over LANs (EAPOL), a protocol that provides a framework for negotiating the authentication method. It defines no explicit authentication protocol itself; the EAPOL is extensible to many authentication protocols. It must be made clear that IEEE 802.1X is not an authentication protocol or a guarantee of a secure authentication algorithm for wireless applications. Authentication protocols like Transport Layer Security (TLS) or Tunneled-TLS (TTLS), among others, can be used with EAPOL.

Some of the terms used in the IEEE 802.1X and its relation with the WLAN are explained below.

- *IEEE 802.1X supplicant:* This is the entity at one end of the point-to-point LAN segment that is being authenticated—the software on the STA that implements the EAP.

- *Authenticator:* This is the entity that facilitates authentication of the entity attached to the other end of that link—the software on the AP that forwards the EAP control packets to the AS, enables/blocks the port, and uses received information.

- *Authentication server:* This is the entity that provides an authentication service to the authenticator, including the radius server, the kerberos server, or the diameter server. It can be integrated into the AP.

The message sequence chart (MSC) of IEEE 802.1X is given in Figure 2.19. The IEEE 802.1X has two ports. The data port at the authenticator is open until the authentication server authenticates the supplicant. Once the supplicant is authenticated, the data port is closed and normal data communication can take place. IEEE 802.1X, together with EAP, allow several different methods of authentication, some of which are mentioned in the figure:

- EAP-TLS;
- EAP-secure remote password (SRP);
- EAP-TTLS;
- EAP-subscriber identity module (SIM) of GSM;
- EAP-authentication and key agreement (AKA) of UMTS;
- EAP-message digest 5 (MD5);
- Protected EAP (PEAP).

2.4.3 IEEE 802.11f

The IEEE 802.11f IAPP is a communication protocol used by one AP to communicate with the other APs. It is part of a communication system comprising the APs, the STAs, a backbone network, and the RADIUS infrastructure [51].

The RADIUS servers provide two functions:

1. Mapping the ID of an AP to its IP address;
2. Distribution of keys to the APs to allow the encryption of the communications between the APs.

The function of the IAPP is to facilitate the creation and maintenance of the wireless network, support the mobility of the STAs, and enable the APs to enforce the requirement of a single association for each STA at a given time.

One of the services the IAPP provides is proactive caching. Proactive caching is a method that supports fast roaming by caching the context of a STA in the APs to which the STA may roam. The next APs are identified dynamically (i.e., without management preconfiguration) by learning the identities of neighboring APs.

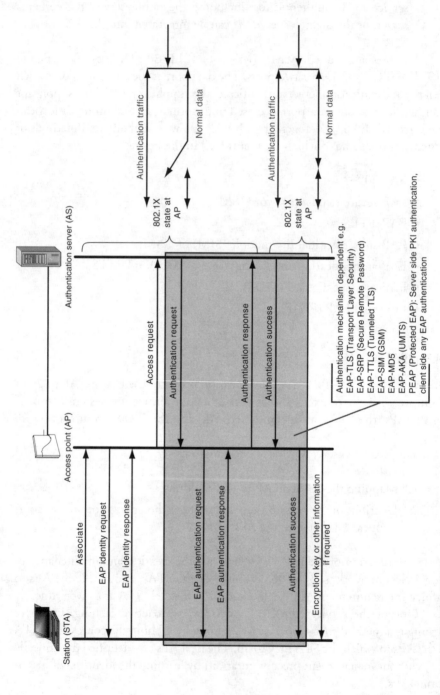

Figure 2.19 IEEE 802.1X EAPOL message sequence chart.

References

[1]   Prasad, R., *Universal Wireless Personal Communications*, Norwood, MA: Artech House, 1998.

[2]   Van Nee, R. D. J., and R. Prasad, *OFDM for Wireless Multimedia Communications*, Norwood, MA: Artech House, 2000.

[3]   Ojanpera, T., and R. Prasad, *Wideband CDMA for Third Generation Mobile Communications*, Norwood, MA: Artech House, 2000.

[4]   GSM MoU, http://www.gsmworld.com.

[5]   ITU, http://www.itu.int/imt.

[6]   The UMTS Forum, http://www.umts-forum.org.

[7]   3GPP, http://www.3gpp.org.

[8]   3GPP2, http://www.3gpp2.org.

[9]   *IEEE Personal Communications*, Vol. 7 No. 2, April 2000.

[10] IEEE 802.11, "Wireless LAN Medium Access Control (MAC) and Physical Layer (PHY) Specifications," November 1997.

[11] IEEE 802.11, "Draft Supplement to Standard for Telecommunications and Information Exchange Between Systems–LAN/MAN Specific Requirements–Part 11: Wireless MAC and PHY Specifications: High Speed Physical Layer in the 5 GHz Band," P802.11a/D6.0, May 1999.

[12] ETSI BRAN, "HIPERLAN Type 2 Functional Specification Part 1—Physical Layer," DTS/BRAN030003-1, June 1999.

[13] HomeRF, http://www.homerf.org.

[14] BLUETOOTH SIG, http://www.bluetooth.com.

[15] *IEEE Personal Communications*, Vol. 7, No. 1, February 2000.

[16] IEEE 802.15, "Part 15.1: Wireless Personal Area Network Medium Access Control (MAC) and Physical Layer (PHY) Specifications," May 2000.

[17] Kamerman, A., and A. R. Prasad, "IEEE 802.11 and HIPERLAN/2 Performance and Applications," ECWT 2000, Paris, October 2–6, 2000.

[18] Prasad, A. R., H. Moelard, and J. Kruys, "Security Architecture for Wireless LANs: Corporate and Public Environment," *VTC 2000 Spring*, Tokyo, May 15–18, 2000, pp. 283–287.

[19] Prasad, A., and A. Raji, "A Proposal for IEEE 802.11e Security," IEEE 802.11e, 00/178, July 2000.

[20] Prasad, A. R., "Performance Comparison of Voice over IEEE 802.11 Schemes," *VTC 1999 Fall*, Amsterdam, September 19–22, 1999, pp. 2636–2640.

[21] Prasad, N. R., et al., "A State-of-the-Art of HIPERLAN/2," *VTC 1999 Fall*, Amsterdam, September 19–22, 1999, pp. 2661–2666.

[22]   Prasad, A. R., "Optimization of Hybrid ARQ for IP Packet Transmission," *International Journal on Wireless Personal Communications*, Vol. 16, No. 3, March 2001, pp. 203–220.

[23]   Prasad, A. R., Y. Shinohara, and K. Seki, "Performance of Hybrid ARQ for IP Packet Transmission on Fading Channel," *IEEE Transactions Vehicular Technology*, Vol. 48, No. 3, May 1999, pp. 900–910.

[24]   Prasad, A. R., and K. Seki, "Capacity Enhancement of Indoor Wireless Communication System with a Novel Channel Sharing Protocol," *ICPWC'97*, Mumbai, India, December 16–19, 1997, pp. 162–166.

[25]   Prasad, A. R., et al., "Wireless LANs Deployment in Practice," in *Wireless Network Deployments*, R. Ganesh and K. Pahelvan (Eds.), Boston, MA: Kluwer Publications, 2000.

[26]   Prasad, A.R., et al., "Performance Evaluation, System Design and Network Deployment of IEEE 802.11," *International Journal on Wireless Personal Communications*, Vol. 19, No. 1, October 2001, pp. 57–79.

[27]   Visser, M. A., and M. El Zarki, "Voice and Data Transmission over an 802.11 Network," *Proc. PIMRC'95*, Toronto, September 1995, pp. 648–652.

[28]   Prasad, N. R., and A.R. Prasad, (Eds.), *WLAN Systems and Wireless IP for Next Generation Communications*, Norwood, MA: Artech House, 2002.

[29]   Prasad, R., *CDMA for Wireless Personal Communications*, Norwood, MA: Artech House, 1996.

[30]   Van As, H. R., "Media Access Techniques: The Evolution Towards Terabit/s LANs and MANs," *Computer Networks and ISDN Systems*, Vol. 26, No. 6–8, 1994, pp. 603–656.

[31]   Chandra, A., V. Gummalla, and J. O. Limb, "Wireless Medium Access Control Protocols," *IEEE Communications Surveys*, http://www.comsoc.org/pubs/surveys, Second Quarter 2000.

[32]   Rom, R., and M. Sidi, *Multiple Access Protocols Performance and Analysis*, New York: Spinger-Verlag, 1990.

[33]   ISO/IEC 8802-11, ANSI/IEEE Std 802.11, First Edition 1999-00-00, Information Technology—Telecommunications and Information Exchange Between Systems—Local and Metropolitan Area Networks—Specific Requirements—Part 11: Wireless LAN Medium Access Control (MAC) and Physical Layer (PHY) Specifications.

[34]   Prasad, A. R., A. Kamerman, and H. Moelard, "IEEE 802.11 Standard," in *WLAN Systems and Wireless IP for Next Generation Communications*, N. R Prasad and A.R. Prasad, (Eds.), Norwood, MA: Artech House, 2002.

[35]   Prasad, N. R., et al., "A State-of-the-Art of HIPERLAN/2," *VTC 1999 Fall*, Amsterdam, September 19–22, 1999, pp. 2661–2666.

[36]   Prasad, N. R., and A. R. Prasad, "Wireless Networking and Internet Standards," in *WLAN Systems and Wireless IP for Next Generation Communications*, N. R. Prasad and A. R. Prasad (Eds.), Norwood, MA: Artech House, 2002.

[37]   Stallings, W., *Cryptography and Network Security: Principles and Practice*, Englewood Cliffs, NJ: Prentice Hall, 1998.

[38] Prasad, A. R., A. Kamerman, and H. Moelard, "IEEE 802.11 Standard," in *WLAN Systems and Wireless IP for Next Generation Communications*, N. R. Prasad and A. R. Prasad (Eds.), Norwood, MA: Artech House, 2002.

[39] Prasad, N. R., and A. R. Prasad, (Eds.), *WLAN Systems and Wireless IP for Next Generation Communications*, Norwood, MA: Artech House, 2002.

[40] Brederveld, L., N. R. Prasad, and A. R. Prasad, "IP Networking for Wireless Networks," in *WLAN Systems and Wireless IP for Next Generation Communications*, N. R. Prasad and A. R. Prasad (Eds.), Norwood, MA: Artech House, 2002.

[41] Bishop, M., *Computer Security: Art and Science*, Reading, MA: Addison Wesley, 2003.

[42] Black, U., *Internet Security Protocols: Protecting IP Traffic*, Englewood Cliffs, NJ: Prentice Hall, 2000.

[43] Borman, D., "Telnet Authentication: Kerberos Version 4," RFC 1411, January 1993.

[44] Kohl, J., and C. Neuman, "The Kerberos Network Authentication Service (V5)," September 1993, Internet RFC: http://www.etf.org/rfc/rfc1510.txt.

[45] Fox, A., and S. D. Gribble, "Security on the Move: Indirect Authentication Using Kerberos," *Proceedings of the Second ACM International Conference on Mobile Computing and Networking (MobiCom '96)*, Rye, New York, November 10–12, 1996.

[46] IEEE 802.11, "Wireless LAN Medium Access Control (MAC) and Physical (PHY) Layer Specifications," ANSI/IEEE, 1999.

[47] Arbaugh, W. A., http://www.cs.umd.edu/~waa/wireless.html.

[48] Rubens, A., et al., "Remote Authentication Dial In User Service (RADIUS)," RFC 2138, April 1997.

[49] IEEE Std 802.1X-2001 "IEEE Standard for Local and Metropolitan Area Networks—Port-Based Network Access Control," June 14, 2001.

[50] Blunk, L., and J. Vollbrecht, "PPP Extensible Authentication Protocol (EAP)," RFC 2284, March 1998.

[51] IEEE P802.11f, "Draft Recommended Practice for Multi-Vendor Access Point Interoperability via an Inter-Access Point Protocol Across Distribution Systems Supporting IEEE 802.11 Operation," D5, January 2003.

[52] Prasad, A. R., H. Wang, and P. Schoo, "Network Operator's Security Requirements on Systems Beyond 3G," *WWRF #9*, Zurich, July 1–2, 2003.

[53] Prasad, A. R., and P. Schoo, "IP Security for Beyond 3G Towards 4G," WWRF #7, Eindhoven, The Netherlands, December 3–4, 2002.

[54] FCC, "Amendment of the Commission's Rules to Provide for Operation of Unlicensed NII Devices in the 5-GHz Frequency Range," Memorandum Opinion and Order, ET Docket No. 96–102, June 24, 1998.

[55] ETSI, "Radio Equipment and Systems, HIgh PErformance Radio Local Area Network (HIPERLAN) Type 1," European Telecommunication Standard, ETS 300–652, October 1996.

[56] Crow, B. P., et al., "IEEE 802.11 Wireless Local Area Networks," *IEEE Communications Magazine*, September 1997, pp. 116–126.

[57] Takanashi, H., and R. van Nee, "Merged Physical Layer Specification for the 5-GHz Band," IEEE P802.11-98/72-r1, March 1998.

[58] ETSI, "Broadband Radio Access Networks (BRAN); HIPERLAN Type 2 Technical Specification Part 1—Physical Layer," DTS/BRAN030003-1, October 1999.

Appendix 2A: ISM Bands

Location	Regulatory Range	Maximum Output Power	Standard
Europe	2,400–2,483.5 MHz	10 mW/MHz (max 100 mW)	IEEE 802.11b,g, HomeRF, WBFH, Bluetooth
	5,150–5,350 MHz	200 mW	
	5,470–5,725 MHz	1,000 mW	HIPERLAN/2
			HIPERACCESS (FWA<11GHz)
			IEEE 802.11a
North America	2,400–2,483.5 MHz	1,000 mW	IEEE 802.11b,g, HomeRF, WBFH, Bluetooth
	5,150–5,250 MHz	2.5 mW/MHz (max. 50 mW)	HIPERLAN Type 2
	5,250–5,350 MHz	12.5 mW/MHz (max 250 mW)	BWIF, IEEE 802.16
	5,725–5,825 MHz	50 mW/MHz (max 1000 mW)	HUMAN IEEE 802.11a
Japan	2,400–2,497 MHz	10 mW/MHz (max 100 mW)	IEEE 802.11b,g, HomeRF, WBFH, Bluetooth
	5,150–5,250 MHz		
	4,900–5,000 MHz (until 2007)	Indoor 200 mW	HIPERLAN Type 2 (MMAC HiSWAN)
	5,030–5,091 MHz (from 2007)		IEEE 802.11a (MMAC)

Appendix 2B: Comparison of WLAN and WPAN Standards

Standard	IEEE 802.11/b	IEEE 802.11a/g	HIPERLAN/2	IEEE 802.15 1.0 and Bluetooth	HomeRF
Mobile Frequency Range (MHz)	2,400–2,483 (North America/ Europe) 2,470–2,499 (Japan)	a: 5,150–5,250 (Europe, North America, Japan) 5,250–5,350 (Europe, North America) 5,470–5,725 (Europe) 5,725–5,825 (North America) 4,900–5,000 (Japan) g: same as IEEE 802.11/b	Same as 802.11a	2,400–2,483 (North America/Europe) 2,470–2,499 (Japan) 2,470–2,499 (Japan)	2,400–2,483 (North America/Europe)
Multiple Access Method	CSMA/CA (distributed and centralized)	CSMA/CA (distributed and centralized)	TDMA (centralized)	TDMA (centralized)	TDMA (Distributed)/ CSMA (Centralized)
Duplex Method	TDD	TDD	TDD	FDD	TDD
Number of Independent Channels	FHSS: 79 DSSS: 3 to 5	a: 12 g: 3 to 5	12	FHSS: 79	FHSS: 79

(continued)

Standard	IEEE 802.11/b	IEEE 802.11a/g	HIPERLAN/2	and Bluetooth	HomeRF
Modulation	FHSS	a/g: OFDM 48 carriers	OFDM 48 carriers	FHSS	FHSS
	GFSK (0.5 Gaussian filter)	6 Mbps BPSK 1/2	6 Mbps BPSK 1/2	GFSK (0.5 Gaussian filter)	GFSK (0.5 Gaussian Filter)
	DSSS	9 Mbps BPSK 3/4	9 Mbps BPSK 3/4		
	DBPSK (1 Mbps), DQSK (2 Mbps)	12 Mbps QPSK 1/2	12 Mbps QPSK 1/2		
	b: DSSS	18 Mbps QPSK 3/4	18 Mbps QPSK 3/4		
	CCK	24 Mbps 16QAM 1/2 36 Mbps 16QAM 3/4 48 Mbps 64QAM 2/3 54 Mbps 64QAM 3/4 g: PBCC and DSSS OFDM optional	24 Mbps 16QAM 1/2 36 Mbps 16QAM 9/16 48 Mbps 64QAM 3/4 54 Mbps 64QAM 3/4		
Channel Bit Rate (Mbps)	1 or 2 b: 5.5 or 11	a/g: 6, 9, 12, 18, 24, 36, 48, and 54	6, 9, 12, 18, 24, 36, 48, and 54	1, 2 or 3	1 or 2

3

Bluetooth: First and Second Generations of WPANs

This chapter provides a technical overview of Bluetooth. After a brief introduction, the Bluetooth architecture is detailed, from both a communication and security perspective. This overview is based on the 2.0 specifications, including the enhanced data rate addenda [1–7].

3.1 Introduction

The initial idea of Bluetooth, originating from an Ericsson project in 1994, was merely to replace wires. Stated like that, the idea seems simple and easy enough to implement. The project evolved and gave birth to the Bluetooth Special Interest Group (SIG), founded by Ericsson, IBM, Intel, Nokia, and Toshiba in 1998. The SIG's mission statement is as follows:

> The goal of Bluetooth is to enable users to connect a wide range of computing and telecommunication devices easy and simply, without the need to buy, carry, or connect cables. It delivers opportunities for rapid ad hoc connections, and the possibility of automatic, unconscious connections between devices. It will virtually eliminate the need to purchase additional or proprietary cabling to connect individual devices.

This mission naturally follows from the original idea, as soon as the real use of cables is taken into account. Moreover, by incorporating the applications in which these cables are used, the definition of the Bluetooth protocol stack evolved into a mix of vertical and horizontal approaches, radically different from the horizontal one of IEEE 802.xx.

The horizontal layered approach [established by the International Standard Organization (ISO)] is based on a *technical viewpoint*. The horizontal approach, which is classical in the networking area, consists of dividing the networking tasks in (seven) layers, each layer being responsible for a family of functional tasks and offering its service to the upper layer. For example, the physical layer is responsible for the transmission and reception of information symbols on the physical medium and offers this service to the upper layer. The WLAN specification bodies—in particular the IEEE 802.11 (Wireless Local Area Network: WLAN) and IEEE 802.15 (Wireless Personal Area Network: WPAN) bodies— actually specify the two lower layers of the ISO protocol stack: the physical layer [Layer 1 (L1)] and the medium access control/logical link control layer [Layer 2 (L2)]. The task of specifying the upper layers is left to the application developers (for WLANs, this makes sense as the upper layers are mostly based on the TCP/IP networking protocol stack, even though specific wireless extensions are desirable).

Bluetooth uses the horizontal ISO layered approach to standardize the physical medium and logical link related functions.

The vertical approach is based on an *application viewpoint*. With this approach, the specification body recognizes that eliminating wires leads to the implementation of a large number of applications. To ease the implementation of these applications on wireless technology, and to ensure interoperability between different vendor's applications, the whole set of layers are specified, from the lower layer (L1) up to the application layer.

Bluetooth uses the vertical approach to tailor each Bluetooth device (or host) to its application (or family of applications), enabling rapid development and ensuring interoperability. This vertical approach gave birth to what is known as *profiles*, each profile being tailored to an application.

The synchronization profile is typical of this vertical approach. The specifications [1] state the following:

> The Synchronization profile defines the requirements for the protocols and procedures that shall be used by the applications providing the Synchronization usage model. This profile makes use of the Generic Object Exchange profile (GOEP) to define the interoperability requirements for the protocols needed by applications. The most common devices using these usage models might be notebook PCs, PDAs (Personal Digital Assistants), and mobile phones.
>
> The scenarios covered by this profile are:
>
> • Usage of a mobile phone or PDA by a computer to exchange PIM (Personal Information Management) data, including necessary log informa-

tion to ensure that the data contained within their respective Object Stores is made identical. Types of the PIM data are, for example, phone-book and calendar items.

- Use of a computer by a mobile phone or PDA to initiate the previous scenario (Sync Command Feature).
- Use of a mobile phone or PDA by a computer to automatically start synchronization when a mobile phone or PDA enters the RF (Radio Frequency) proximity of the computer.

This description of the scope of the synchronization profile shows us that a profile defines all protocols and procedures for a specific application (also named usage model) and can be based on another profile. In the case of the synchronization profile, it is based on the GOEP, which is itself based on the serial port profile. Figure 3.1 illustrates the different layers used for the serial port profile.

The serial port profile is based on the different layers that are sketched in Figure 3.1. It specifies parameters used in these protocols as well as the procedures used (e.g., to establish links), and also how these procedures use the underlying protocols.

In this profile, the baseband, Link Manager Protocol (LMP) and Logical Link Control and Adaptation Protocol (L2CAP) are the OSI layer 1 and 2 Bluetooth protocols. RFCOMM is the Bluetooth adaptation of GSM TS 07.10, [7] providing a transport protocol for serial port emulation. SDP is the Bluetooth Service Discovery Protocol. The port emulation layer shown in Figure 3.1 is the entity emulating the serial port, or providing an application programming interface (API) to applications. The applications on both sides are typically legacy applications, able and wanting to communicate over a serial cable (which in this case is emulated). But legacy applications cannot know about Bluetooth proce-

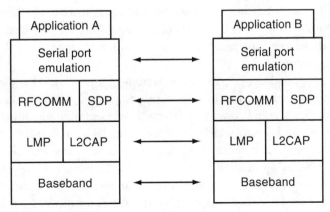

Figure 3.1 Protocol stack of the serial port profile.

dures for setting up emulated serial cables, which is why they need help from some sort of Bluetooth-aware helper application on both sides.

At the lower layers—as Bluetooth is meant to be a low-power, low-cost, low-complexity solution—it offers a rather low data rate (maximum data rate in one direction of 721 Kbps and 2.1 Mbps in the enhanced rate mode) in the ISM 2.4-GHz band. The networking layer is based on a simple star architecture (the piconet) and on the interconnection of these piconets to form a scatternet, giving birth to simple and small ad hoc networks but allowing the creation of large (and possibly complex) ad hoc networks.

3.2   Technical Overview of Bluetooth 2.0

The Bluetooth specification [1] is divided into 13 volumes, three of which are currently published. The first five of them correspond to the nonapplication-specific parts: the architecture definition (Vol. 1), the core system (Vols. 2 and 3), the host controller interface (HCI) (Vol. 4), and the SDP (Vol. 5). The remaining volumes specify the application-specific profiles. Note that the three first volumes contain 1,200 pages, the other volumes are currently borrowed from earlier specifications and addenda to these specifications.

We will mainly follow the core/profile division in this section, giving an overview of the architecture and protocol stack, as well as security operation and core enhancements.

The Bluetooth protocol stack is depicted in Figure 3.2. As stated before, this protocol stack differs from the OSI 7-layer protocol architecture in the sense that the horizontal layering principle is violated.

The lower part of the Bluetooth protocol stack consists of the radio and baseband layers, as well as the link manager, and corresponds to the two lower ISO layers (hence following a classical horizontal approach). The upper part of the Bluetooth protocol stack consists of a large collection of (mostly existing, some slightly adapted) preexisting protocols, which are tailored to the applications to be supported.

The lower and upper parts are separated by the HCI. The lower part is meant to be implemented in the Bluetooth device (the baseband in hardware and the link manager in firmware); whereas the upper layers are meant to be implemented in the host to which the Bluetooth device is attached (a mobile phone or computer) and hence are implemented in software. The HCI provides a uniform interfacing method for accessing the Bluetooth hardware capabilities. The HCI specifications are implemented in an HCI driver, which is independent from the HCI transport method (this transport method can be USB or PC card).

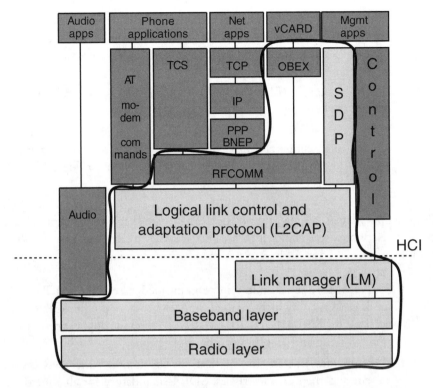

Figure 3.2 A subset of the Bluetooth protocol stack.

The protocols shown in the light gray boxes are mandatory for every Bluetooth application (Figure 3.2). The bubble illustrates the GOEP, comprising all protocols that are needed to implement this profile.

Before detailing the operation of Bluetooth, we provide a quick overview of the most important protocols.

Lower Part

- The radio layer operates in the 2.4-GHz ISM band, employing frequency hopping spread spectrum, on 79 channels (each with 1-MHz bandwidth) with a symbol rate of 1 Mega symbol per second (Msps) and a Gaussian frequency shift keying (GFSK) modulation, leading to a bit rate of 1 Mbps. The enhanced rate specifications introduce the use of π/4-DQPSK (differential quaternary phase shift keying) and 8-DPSK, leading to bit rates of 2 and 3 Mbps.
- The baseband layer (link controller) specification defines the multiple access and duplexing method used (TDMA and TDD, respectively).

It also defines the various components of the channels used (physical channels, transport channels, logical links and link control) and the way they interact—and hence how the networks are formed (i.e., the piconet).

- The link manager (LM) defines the control oriented logical links that control the baseband and radio layers.

- L2CAP defines channel-based abstraction to applications and services, carrying out segmentation/reassembly of application data and multiplexing/demultiplexing of L2CAP channels over (lower) logical links.

Upper Part

For the upper layers, all relevant protocols that serve specific applications (e.g., cordless telephony, serial links, local area networking, and so forth) are specified. Sample protocols shown in Figure 3.2 include the following:

- RFCOMM is a serial cable emulation protocol based on ETSI TS 07.10 specifications [7] (which are used for GSM and are the European specifications for RS-232). RFCOMM serves for phone applications as well as networking and object exchange applications.

- SDP provides a means for applications to discover which services are present and their characteristics. SDP is mandatory for all Bluetooth-based applications.

- For networking applications, the TCP/UDP and TCP/IP protocols are used, as well as the PPP/Bluetooth Network Encapsulation Protocol (BNEP), for connection through a serial link.

- For phone applications, the Telephony Control Protocol Specification (TCS) offers cordless phone operation. The use of AT modem commands is another way to support phone-based networking operation.

- For object exchange applications, such as vCard (an electronic business card application), the Object Exchange (OBEX) protocol (which is an adaptation of the irOBEX protocol for infrared devices) is specified.

- Audio applications are served by a specific audio protocol based on continuously variable slope delta (CVSD) coding, which is based on a 64-Kbps synchronous channel.

Profiles

The core specifications define a generic access profile (GAP), from which all other profiles are derived, with specific additions. GAP defines the procedures for establishing links between Bluetooth devices, the general security architecture and modes, as well as discoverability modes (whether a module can be discovered).

There are more than 20 specific profiles (which we will not detail here), including networking applications (line access profile, file transfer profile), phone applications (cordless, fax, dial-up networking profiles), object push/pull based profiles (synchronization profile), and audio (headset profile). As Bluetooth is becoming more and more pervasive in everyday life, many new profiles are being developed.

3.3 Core Specification

The Bluetooth core system covers the four lowest layers defined above, as well as the SDP and the GAP. It is based on an underlying transport architecture depicted in Figure 3.3, which follows more or less the OSI taxonomy defining the

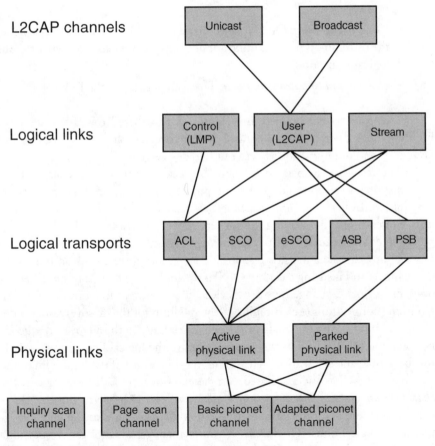

Figure 3.3 Bluetooth data transport architecture. (*From:* [1].)

physical layer, the logical layer, and the L2CAP layer, which is in Layer 2 of the OSI model.

The transport architecture (Figure 3.3) is based on the following:

- *Four physical channels:* The inquiry and page scan channels (used for the discovery procedures), the basic piconet channel (which was the only data channel in Bluetooth 1.1), and the adapted piconet channel, using a form of link adaptation to enhance performance in situations where the channel undergoes a lot of interference and/or noise.

- *Two physical links:* The active physical link is the normal link, whereas the parked physical link enables a piconet member to go in a parked mode, for low power operation (and also to allow a piconet to address a total of 255 slaves, whereas only seven active slaves are allowed).

- *Five logical transports:* Some support data and voice channels for the active channels, and one is a transport for the parked link (these are detailed below).

- *Three logical links:* One for the control plane and one for the user plane; the third link is for streaming applications (hence based on synchronous logical transports).

- *Unicast and broadcast channels:* These are located at the link layer.

The core system architecture (Figure 3.4) is based on rather standard building blocks, with user and control paths between each layer in a device and layer-to-layer communications between two devices.

The *RF block*, associated with the physical channels, is in charge of the transmission and reception of packets on physical channel, and it is controlled by the link controller. The *link controller*, associated to the physical links, is in charge of coding and decoding the packets, as well as flow control, acknowledgments, and ARQ management. The *baseband resource manager*, associated to the physical links and channels, is responsible for granting time slots on the physical channels and insuring a certain QoS to the higher layers. It is responsible not only for normal exchange between devices in connected mode, but also for establishing connections (i.e., listening to the medium for the discovery of new devices) and monitoring the quality of the channels (e.g., in the adaptive frequency hopping mode). The *link manager*, associated to the logical links, is responsible for the management (creation, release, and modification) of the logical links and transports as well as the associated parameters. On top of all these blocks, the *channel manager* and *L2CAP resource managers* are in charge of L2CAP channel management and interaction with the link manager for creation and configuration of logical links. The L2CAP resource manager block is also responsible for scheduling between channels to ensure that L2CAP channels with QoS

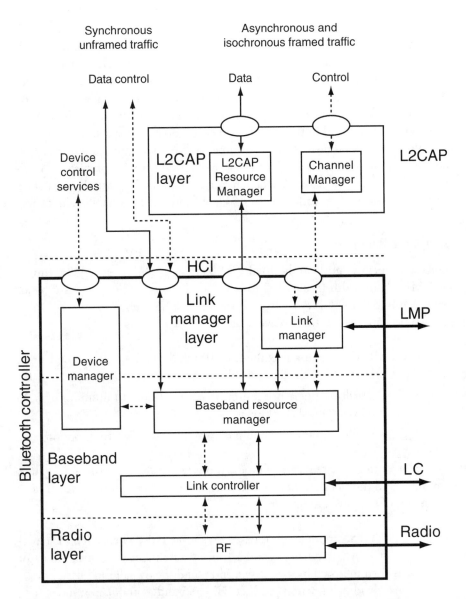

Figure 3.4 Bluetooth core system architecture. (*From* [1].)

commitments (e.g., a channel with a frame error rate less than 10^{-6} and a maximum delay of 200 ms) are not denied access to the physical channel due to Bluetooth controller resource exhaustion. This is required because the architectural model does not assume that the Bluetooth controller has limitless buffering, or that the HCI is a pipe of infinite bandwidth. L2CAP resource managers may also carry out traffic conformance policing to ensure that applications are submitting data to the L2CAP within the bounds of their negotiated QoS settings. The general Bluetooth data transport model assumes well-behaved applications and does not define how an implementation is expected to deal with this problem. The *device manager* is not associated to data transport, but it is in charge of the general behavior of the device (network knowledge, identification, discoverability, link keys).

3.3.1 The Radio Layer

Bluetooth operates in the unlicensed ISM band at 2.4 GHz, using 79 channels of 1-MHz bandwidth from 2.402 to 2.480 GHz, with additional guard bands from 2.400 to 2.402 GHz and from 2.480 to 2.483.5 GHz.[1]

Bluetooth defines three power classes:

- Class 1 will output a maximum of 100 mW and a minimum of 1 mW. This class of devices will be capable of power control, in steps of 2 to 8 dB.

- Class 2 will output a maximum of 2.5 mW and a minimum of 0.25 mW; power control is optional.

- Class 3 will output a maximum of 1 mW; power control is optional.

Bluetooth uses three types of modulations whose main advantages are their simplicity and the fact that they show a (quasi-)constant amplitude, which eases the design of the power amplifier.

1. The GFSK modulation is based on simple FSK, where a 0 bit is represented by a frequency f_0 and a 1 bit by a frequency f_1. This modulator is then followed by a Gaussian filter, and the overall modulation has a bandwidth-bit period product BT = 0.5 and a modulation index (i.e., the ratio between the frequency deviation and the modulating frequency) between 0.28 and 0.35. Note that using simple FSK means that one modulates 1 bit per symbol, which leads, for 1 Msps, to 1 Mbps.

2. The $\pi/4$-DQPSK modulation is based on differential quaternary phase shift keying. In QPSK, 2 bits are grouped together and each group of

---

1. National regulations result in slight differences from country to country.

bits is coded in four phases (e.g., $00 \rightarrow 0$ and $01 \rightarrow \pi/2$). To be closer to a constant amplitude signal, the symbols are $\pi/4$ rotated from one time symbol to the other and differential encoding is performed for robustness (to channel impairment) purposes. This leads to a bit rate of 2 Mbps, which is introduced in the enhanced data rate extension to Bluetooth 2.0

3. The 8-DQPSK is based on 8 PSK and also uses differential encoding. As 3 bits are used to form the eight different phases, this modulation offers a 3-Mbps rate to Bluetooth 2.0 in its enhanced data rate mode.

On the receiver side, the most striking characteristic of Bluetooth is its rather low sensitivity level [defined as the level for which a raw bit error rate (BER) of 0.1% is met for 1-Mbps operation and 0.01% for 2 and 3 Mbps], which is equal to –70 dBm for any Bluetooth receiver. This quite low sensitivity is meant to enable the design of low-cost Bluetooth receivers.

3.3.2 The Baseband Layer

The baseband layer of Bluetooth defines the basic Bluetooth networks: the piconet and the scatternet (Figure 3.5). A piconet is a collection of two or more Bluetooth devices sharing the same physical channel (i.e., the same frequency hopping sequence; see below). A piconet is formed by a master and up to seven active slaves. The restriction to seven slaves, taken to enable decent transmission rates on the channels, is somewhat alleviated by the possibility of having about 250 more parked slaves, which are not in communication with the master but can be brought back to an active state in a short time. A scatternet is a collection of interconnected piconets.

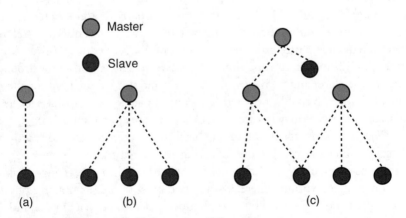

Figure 3.5 Simple Bluetooth (a, b) piconets and (c) a scatternet.

The basic building block of the baseband is the *physical channel*. Physical channels are defined by a pseudo-random code that drives the frequency hopping (FH) process. The hopping rate is 1,600 hops/s in the connection mode and 3,200 hops/s in the inquiry and page modes.

Bluetooth defines four physical channels: basic piconet, adapted piconet, page scan, and inquiry scan physical channels.

1. The basic piconet channel is used for communication between connected devices during normal operation.

2. The adapted piconet channel offers the possibility of using a subset (minimum 20) of the 79 frequency channels, allowing interference avoidance (i.e., avoid frequencies where the interference level is too high to have decent performance) and easing coexistence with other systems in the same band. Moreover, in the adapted piconet channel, the slave uses the same frequency as the master in its preceding transmission—this is slightly different than the normal operation depicted in Figure 3.6; namely, the frequency used in $f(k+1)$ is the same as the frequency used at $f(k)$.

3. The paging scan channel is used to page a connectable device (one that is prepared to accept connections) and to initiate the connection procedure. The average delay before the master reaches the slave is about half a second, with a maximum of 2.56 seconds.

4. The inquiry scan channel is used in inquiry mode in order for a device to be discovered (including public printers and fax machines with an unknown address).

Without going too far into the technical details of the FH-sequence choice and in the synchronization of the piconet, it suffices to say that the physical channels are based on a TDD scheme, with each party of a link sending information in its turn, and on time slotting, where in each slot of duration 625 µs a different frequency is used and a packet is sent from a sender to a receiver. Moreover, each odd-numbered time slot will be used to send information from the master to the slave, and each even-numbered time slot will be used for the opposite direction. To enable more efficient communications, multislot packets are also used; one packet may last for 1, 3, or 5 time slots (see Figure 3.6).

Physical links are the next building block. They are associated to exactly one physical channel and provide the following additional features: power control, multislot packet control, and channel quality-driven data rate change, link supervision, and encryption. There are only two types of links: the active physical link and the parked physical link, the latter being dedicated to a physical link between a master and a parked slave.

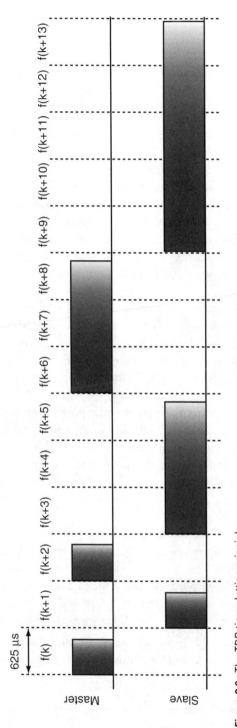

Figure 3.6 The TDD time slotting principle.

There are five *logical transports*:

1. The asynchronous connection-oriented (ACL) logical transport;
2. The synchronous connection-oriented (SCO) logical transport;
3. The extended SCO (eSCO) logical transport;
4. The active slave broadcast (ASB) logical transport;
5. The parked slave broadcast (PSB) logical transport.

By default, Bluetooth connection creates an ACL transport on an active physical link for asynchronous data transfer. ACL can also be used to carry isochronous framed data (e.g., VoIP). The data transmission employs an automatic request (ARQ)[2] scheme to ensure the communication reliability by using CRC for error detection and retransmission upon occurrence of transmission error.

The master schedules the ACL transmission in the piconet by allocating transmission slots to the slaves. SCO links have priority over ACL, so ACLs can claim only unused slots.

The SCO transport is a symmetric synchronous transport, basically used for synchronous operation—for example, voice communications on a streaming link coming from the upper layers. The SCO reserves slots on the physical channel and is targeted for 64-Kbps stream communications, as it is mainly designed to transport voice using 64-Kbps CVSD modulation or pulse coded modulation (PCM).

The eSCO is a symmetric or asymmetric point-to-point link. The eSCO also reserves slots on the physical layers, but offers additional flexibility to combine packet types, allowing for a range of synchronous bit rates to be supported. eSCO can also offer limited retransmission of packets and also allow other connections to steal some slots, if they are not used for retransmission.

The ASB and PSB are used for user/control communication to the whole piconet and are unreliable.

Table 3.1 gives an overview of the data rates and communication types that can be used. The first column of this table gives the type of packet, where:

- DMx and DHx, respectively, stand for data medium and data high, and x is the number of slots occupied by the packet. DMx packets are protected by 2/3 FEC codes, while DHx packets are not protected.
- HV1, HV2, and HV3 packets transport 10, 20, and 30 bytes, respectively, on an SCO link. HV stands for high-quality voice.

---

2. The sender waits for an acknowledgment from the receiver; if the acknowledgment is not received, the sender retransmits the packet.

- DV stands for data-voice. These packets transport a mix of data and voice, supporting one voice channel and one synchronous data channel at 57.6 Kbps with 2/3 FEC protection.
- EV3, EV4, and EV5 packets transport up to 30, 120, and 180 information bytes, respectively. EV4 is protected by a rate 2/3 FEC. EV stands for extended voice.
- Prefix 2 and 3 stand for 2 and 3 Mbps on the air rate.

Table 3.1
Transport Packets and Data Rates

Type	Over-the-Air Rate (Mbps)	FEC	CRC	Symmetric Max Rate (Kbps)	Asymmetric Max Rate (Kbps)	
					Forward	Reverse
DM1	1	2/3	Yes	108.8	108.8	108.8
DH1	1	–	Yes	172.8	172.8	172.8
DM3	1	2/3	Yes	258.1	387.2	54.4
DH3	1	–	Yes	390.4	585.6	86.4
DM5	1	2/3	Yes	286.7	477.8	36.3
DH5	1	–	Yes	433.9	723.2	57.6
AUX1	1	–	–	185.6	185.6	185.6
2-DH1	2	–	Yes	345.6	345.6	345.6
2-DH3	2	–	Yes	782.9	1,174.4	172.8
2-DH5	2	–	Yes	869.1	1,448.5	115.2
3-DH1	3	–	Yes	531.2	531.2	531.2
3-DH3	3	–	Yes	1,177.6	1,766.4	265.6
3-DH5	3	–	Yes	1,306.9	2,178.1	177.1
HV1	1	1/3	–	64.0		
HV2	1	2/3	–	64.0		
HV3	1	–	–	64.0		
DV	1	2/3 D	Yes D	64.0+57.6D		
EV3	1	–	Yes	96.0		
EV4	1	2/3	Yes	192.0		
EV5	1	–	Yes	288.0		
2-EV3	2	–	Yes	192.0		
2-EV5	2	–	Yes	576.0		
3-EV3	3	–	Yes	288.0		
3-EV5	3	–	Yes	864.0		

Logical links between devices are either framed (carried by L2CAP channels on the user plane and carried by LMP channels on the control plane) or un-framed (to carry streaming audio/video).

Connection Procedures and Low-Power Modes

Before connection and/or to acquire knowledge about the environment, the Bluetooth device may use an inquiry procedure to discover nearby devices or to be discovered. Bluetooth devices can be *discoverable*, hence responding to the inquiry procedure or refusing to respond. Once the Bluetooth device knows of a possible Bluetooth mate, it can start a paging (connection) procedure, which will be successful only if the Bluetooth mate is connectable (i.e., listening to the paging channel). If the connection procedure is successful, the devices enter in the connected mode. They form (or are part of) a piconet and share an ACL link. When in connected mode they can create additional ACL links and/or SCO-eSCO links.

Bluetooth offers different low-power modes for improving battery life. Piconets are formed on demand when communication among devices is ready to take place. At all other times, devices can be either turned off or programmed to wake up periodically to send or receive inquiry messages. It is possible to switch a slave into a low-power mode whereby it sleeps most of the time and wakes up only periodically. Three types of low-power modes have been defined:

- In the *hold mode*, the physical link is only active during slots that are reserved for the operation of synchronous links, thus disallowing all asynchronous links.

- *Sniff mode* is used to put a slave in a low-duty cycle mode, whereby it wakes up periodically to communicate with the master or to engage activity on another physical channel. The sniff mode only applies to the default ACL logical transport and does not affect SCO or eSCO logical transports.

  The Sniff and Hold periods are programmable and can range from 1.25 ms to 40.9 sec, allowing very low-duty cycles if needed.

- The *parked state* is similar to the sniff mode, but it is used to stay synchronized with the master without being an active member of the piconet, hence disallowing any logical link (except PSB, which is used for communication between the piconet master and the parked slave). The parked mode enables the master to admit more than seven slaves in its piconet. The parked periods are programmable and can range from 8.75 ms to 40.9 sec.

3.3.3 Host Controller Interface

The HCI, a major part of the specifications (about 25%, or 300 out of 1,200 pages) is a significant helper for the partitioning of Bluetooth between the host and the device, as well as for portioning between hardware and software (Figure 3.7). The main roles of the HCI are to:

- Provide an independent implementation of hardware;
- Provide a standard interface to the link manager;
- Provide access over standard transport layers (USB/UART) between the Bluetooth module and the host controller.

The HCI is architectured on three main modules:

- The HCI firmware, located on the host controller (most probably on the Bluetooth device), implements the HCI commands for the Bluetooth hardware by accessing the (programming) resources of the lower layers.
- The host controller transport layer is responsible for transferring the HCI commands/data on specific transport layers (initially USB – UART – RS232).
- The HCI driver, located on the host (PC-mobile phone), provides the interface between the HCI firmware (through the host controller transport layer) and the host (i.e., the higher layers of Bluetooth that are implemented on the host).

From a software perspective, Bluetooth HCI is based on *commands*, which carry the information and the control data units; and on *events*, which are responsible for the control of the HCI operations, including error reporting and flow control on the interface.

3.3.4 Logical Link Control and Adaptation Protocol

The L2CAP is part of the data link layer and, through the HCI, talks to the baseband layer (link controller) and to the upper layers (network/application) (see Figures 3.8 to 3.10).

L2CAP provides:

- L2CAP (logical) channels, mapped to L2CAP logical links, mapped on ACL logical transports (see Figure 3.3);
- Up to 64-kB service data units (i.e., it accepts 64-kB packets of data from the upper layer);
- Per-channel flow control and retransmission.

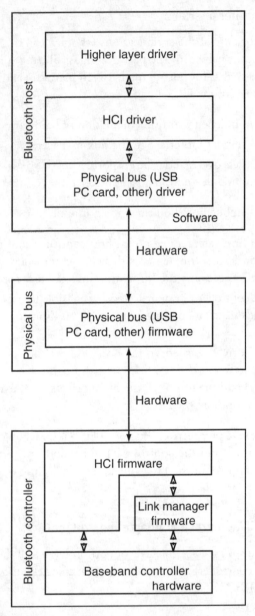

Figure 3.7 The HCI helps the design of the software/hardware partitioning and the Bluetooth device/Bluetooth host partitioning.

Figures 3.8, 3.9, and 3.10 illustrate how L2CAP communicates with its peers (Figure 3.8), which upper and lower layers are used (Figure 3.9), and what its functions are (Figure 3.10).

Without fully going into the technicalities of the L2CAP, its main features and assumptions can be summarized as follows.

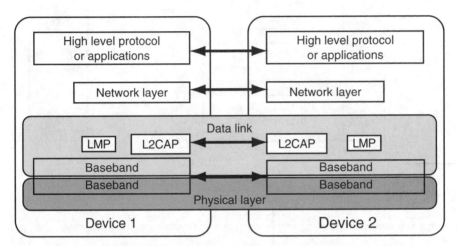

Figure 3.8 L2CAP in the protocol stack. L2CAP is above the baseband layer and complements the link manager to form the data link layer. It receives data from the upper layer, which is mainly a network layer.

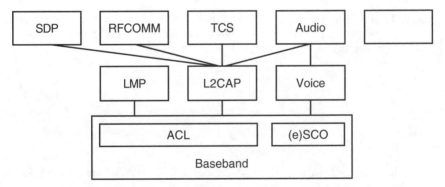

Figure 3.9 The upper layers for L2CAP are the SDP, the RFCOM protocol (the emulation of serial line protocols), the phone-oriented protocol, and the audio protocol. It relies on ACL transport channels (i.e., it is data oriented).

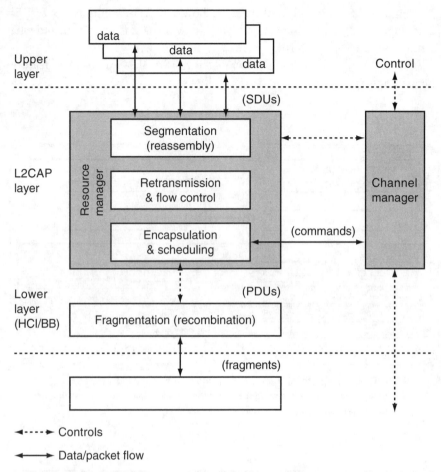

Figure 3.10 L2CAP is mainly responsible on the data plane for segmentation for higher layers, and retransmission and flow control and fragmentation/recombination for lower layers, and for channel management on the control plane.

Assumptions

- L2CAP uses the underlying ACL logical transport, where ordered delivery of data packets is assumed, although packet corruption and/or packet duplication may occur.
- Baseband provides the impression of full-duplex communication channels, even if unicast/multicast traffic only requires simplex (data) channels.

Features

L2CAP provides asynchronous connection-oriented and connectionless (for broadcast operation) channels, with some QoS support. L2CAP does not transport audio or synchronous data (which is transported over SCO or eSCO logical transport channels), although VoIP would be carried by L2CAP. L2CAP may operate in three different modes, selected by the upper layers for each L2CAP channel.

1. *Basic L2CAP mode*, which is the default mode, is used when no other mode is agreed upon. In this mode, only basic segmentation, protocol multiplexing, encapsulation, scheduling, and fragmentation are performed.
   - *Protocol/channel multiplexing:* L2CAP supports multiplexing, allowing each peer device to route the connection request and data units to the correct upper layer protocol, as the baseband is not aware of the type of data transmitted.
   - *Segmentation and reassembly:* The length of the transport frames is controlled by the individual applications running over L2CAP. Many multiplexed applications are better served if L2CAP has control over the length of the packets it handles. This provides the following benefits:
     1. Segmentation will allow the interleaving of application data units in order to satisfy latency requirements.
     2. Memory and buffer management is easier when L2CAP controls the packet size.
     3. Error correction by retransmission can be made more efficient.
     4. The amount of data that is destroyed when an L2CAP PDU is corrupted or lost can be made smaller than the application's data unit.
     5. The application is decoupled from the segmentation required to map the application packets into the lower layer packets.
   - *Basic flow control/acknowledgement:* This is based on the 1-bit sequence number of the link controller (except for baseband broadcast packets).
   - *Fragmentation and reassembly:* This is similar to segmentation, but with the goal of adapting to the lower layers (i.e., to their limited capacity, or their need to send small packets in a noisy channel), the L2CAP will fragment its packets before handing them over to the lower layers and also take care of reassembly when receiving packets from the lower layers.

2. In *flow control mode,* the PDUs are numbered and acknowledged. The sequence numbers in the PDUs are used to control buffering, and a TxWindow size (of maximum 31) is used to limit the buffer space and to provide flow control. Buffer size can also be controlled. In flow control mode, missing PDUs are detected but not retransmitted.

3. *Retransmission mode* is similar to the flow control mode, but it ensures that all PDUs are delivered (within certain time limits), hence allowing some QoS guarantees. Retransmission is based on a simple go-back-n protocol, which limits the buffering requirements but is less bandwidth efficient.

Note that flow control and retransmission modes are based on a per L2CAP channel operation, allowing it to offer different flow control/retransmission characteristics to different applications. By resorting to these three modes, L2CAP can apply two different types of QoS to applications, similar to RFC 1363 and specified for the two (Tx and Rx) directions:

- Best effort;
- Guaranteed QoS (including average rate and peak rate, to allow for bursty communications, and average delay and delay variation, for IP telephony applications).

3.3.5   Higher Layer Protocols

The main higher layer protocols are as follows (see Figure 3.2):

1. *Bluetooth Network Encapsulation Protocol.* This is used by the PAN profile. The role of BNEP is to offer support for IP-based networking protocols (including IPv6). BNEP has to be able to support other popular protocols other than those that are IP-based and come with a low overhead. In this context, Bluetooth is considered to be a transmission medium in the same OSI layer as Ethernet (i.e., Layer 1 and the lower part of Layer 2). The packet format is similar to that of Ethernet packets.

2. *Telephony Control Protocol Specification–Binary.* This is used for telephony applications and provides call control, group management, and connectionless TCS (to exchange signaling information not related to an ongoing call). Note that TCS is able to provide point-to-multipoint communications and, besides the use of SCO channels for voice, resort to ACL channels for call control and group management.

3. *RFCOMM.* This provides serial port emulation over the L2CAP protocol, based on ETSI standard TS 07.10, and it can emulate up to 60

connections between two Bluetooth devices. RFCOMM is a basis for the network-oriented protocols (BNEP/PPP) and OBEX.

4. *Audio.* This is based either on PCM coding or on CVSD coding, on a 64-Kbps link and can be protected by FEC, resulting on a good toll quality voice communication.

5. *Object Exchange Protocol.* This is defined by infrared data association (IrDA) to interconnect the full range of devices that support IrDA protocols. The main advantage of OBEX is that it supports many higher layer application protocols for synchronization (vCard and vCalendar), file transfer, and object exchange (push/pull).

6. *Audio Video Distribution (Control) Transfer Protocol (AVDTP and AVCTP).* This is for point-to-point audio/video streaming media. The transfer is done over L2CAP channels, taking benefit of the pseudo-isochronous feature of L2CAP channels.

3.3.6   Service Discovery Protocol

The SDP enables a Bluetooth device (possibly with limited capabilities) to inquire (and to provide) what services are available in its environment as well as the detailed characteristics of the service, which can be provided incrementally to minimize bandwidth usage if the service is not needed. A Bluetooth device can either browse the services in an unstructured manner, or take advantage of the structure of services (in programming terms, the services are represented by classes; in functional terms, a class may comprise, for example, a family of printers being a subclass of the printing class). Each Bluetooth device will maintain a service database that will be accessible to the neighbors, enabling the publishing of other devices' services.

As the number of services is expected to increase (possibly explode), SDP provides procedures to sort the variety of services and to identify how unknown services work. It will also provide for the creation and definition of new services without requiring registration with a central authority.

SDP is responsible only for locating the services; access to the services is out of its scope. Note also that services are *fixed*; that is, there is no support for negotiation of parameters and for control/change of operation. It is worth noting that SDP is based on a pull paradigm (like that illustrated in Figure 3.11, the client has to pull the information from the server by issuing requests) and does not provide event notifications when services are modified or become unavailable. The SDP protocol is based on a simple client/server model, and the protocol consists of SDP requests and SDP responses, as illustrated in Figure 3.11.

The SDP server maintains a list of service records. Each service record contains a list of attributes that characterize the services associated with the server.

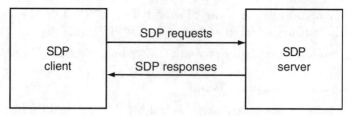

Figure 3.11 The basic SDP client/server protocol.

Each Bluetooth device can have a maximum of one SDP server and SDP client, each application notifying the SDP server of the services it provides and querying the client SDP to query services on remote servers. Note that because of the ad hoc nature of the network, the pool of SDP clients/servers can change in time; and SDP provides the procedures to cope with these changes.

3.4 Profiles

Interoperability between devices from different manufacturers is essential for Bluetooth technology to succeed. A profile defines a selection of messages and procedures (generally termed capabilities) from the Bluetooth SIG specifications and gives an unambiguous description of the air interface for the specified service. The Bluetooth certification authority uses the profiles to test and certify compliance, and it grants use of the Bluetooth logo only to products that conform to the methods and procedures defined in the profiles.

3.4.1 Profile Architecture

Version 2.0 of the specification provides nearly 30 profiles (some in the pre-1.0 version). GAP (the only profile that is part of the core specification) defines generic procedures for discovery and the link management aspects of connecting Bluetooth devices. All Bluetooth devices must implement this profile. Additional profiles are then derived from GAP and structured according to Figure 3.12, where only a subset of the most relevant profiles are drawn.

3.4.2 Generic Access Profile

According to the Bluetooth Specifications [1], GAP "defines the generic procedures related to discovery of Bluetooth devices (idle mode procedures) and link management aspects of connecting to Bluetooth devices (connecting mode procedures). It also defines procedures related to use of different security levels. In

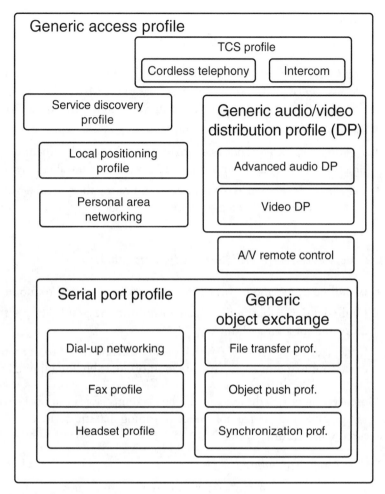

Figure 3.12 Profile architecture.

addition, this profile includes common format requirements for parameters accessible on the user interface level."

It is interesting to dig into the details of GAP, as it defines the overall operation of Bluetooth and defines its main modes. GAP describes the use of the lower layers of the Bluetooth protocol stack (baseband, link manager protocol, and L2CAP).One of the most important aspects of GAP is the definition of the security mode (detailed below in the Section 3.5), which is the reason why GAP also includes the use of higher layer protocols (SDP, TCS, RFCOMM, and OBEX).

3.4.2.1   Interface Level

The user of a Bluetooth device can be in the presence of:

- *The Bluetooth device address* (BD_ADDR), which is the baseband level address, coded on 48 bits, and very similar to the MAC address, with the first 3 bytes being vendor specific and the other three being device specific;
- *The Bluetooth device name*, coded on 248 bytes maximum;
- *The Bluetooth passkey (Bluetooth PIN)*, which is used to authenticate two Bluetooth devices and can be up to 128 bits (16 bytes) long;
- *The class of device*, which is a parameter indicating the type of device and the type of services that are supported by this device.

3.4.2.2   Modes

Bluetooth devices can be in the following discovery modes:

- Nondiscoverable mode, which means it will not respond to inquiry;
- Limited discoverable mode, which means it will respond to a specific inquiry and for a limited time;
- General discoverable mode, which means it shall respond to a general inquiry (not to a limited access code inquiry), possibly for a limited time.

Moreover, Bluetooth devices can also be in connectable or nonconnectable mode, or in pairable and nonpairable mode (meaning that the device can or cannot be paired when security is enabled).

3.4.2.3   Idle Mode Procedures

The initiator of a Bluetooth connection can use five procedures.

1. *General inquiry:* Provides the initiator with the BD_ADDR, clock, and class of device of the general or limited discoverable devices;
2. *Limited inquiry:* Provides the same information, but only for limited discoverable devices;
3. *Name discovery:* Provides the device name of connectable devices;
4. *Device discovery:* Provides the initiator with the BD_ADDR, clock, class of device, and device name of discoverable devices (i.e., this is a general or limited inquiry followed by a name discovery);
5. *Bonding:* Creates a relation between two Bluetooth devices based on a common link key (a bond). The bonding procedure is responsible for the creation and exchange of the key. In addition to pairing, bonding can involve higher layer initialization procedures.

3.4.2.4 Establishment Procedures

After discovery, three different procedures can take place.

1. *Link establishment:* Establishes a physical link (default ACL) between two Bluetooth devices using the baseband and LMP functionalities. This includes the security operation of authentication and encryption if needed.

2. *Channel establishment:* Establishes a Bluetooth logical link between devices using the L2CAP layer.

3. *Connection establishment:* Establishes a connection between applications on two Bluetooth devices.

3.5 Security

Security issues are important in wireless technologies since a signal can be listened to by anybody and it is quite easy to eavesdrop on the communication by sending unwanted signals. Security is even more crucial in Bluetooth, as a security gap gives access to the applications and to personal data directly, rather than merely to the network.

As indicated in Chapter 8, there is an arsenal of techniques to secure the links, ranging from application-oriented protocols to data link methods. Bluetooth basically uses classical link methods to secure its communications, allowing higher layer security to come on top of it. On the whole, one may say that Bluetooth is reasonably secure; that is, if you are aware of security issues and follow the general recommendations of the specifications.

3.5.1 General Operation

The general security operation is defined by GAP. According to GAP, Bluetooth devices can be in one of three security modes:

- Mode 1: no security. This is to be avoided.
- Mode 2: service level enforced security. In this mode, security can be enforced after physical link establishment and before L2CAP channel establishment. Security will be enforced depending on the requirements of the requested channel or service. These (service level) requirements will be classified as:
- Authorization required;
- Authentication required;
- Encryption required.

- Mode 3: link level enforced security. In this mode, security shall be enforced before physical link establishment, thus disallowing the connection with untrusted devices.

Additionally, for mode 2, one can define a *trusted device*, which will have unrestricted access to all of the host device services, and an *untrusted device*, which will have no trusted fixed relationship and thus limited access to services (though services with no specific security requirements can be granted).

3.5.2 Link Level Enforced Security

Link level security is managed by the Link Manager Protocol and is based on keys serving for authentication and encryption. Besides the PIN code (from 1 to 16 bytes), which is used for key generation and authentication, the four entities that are used at the link layer are:

1. The Bluetooth Address (BD_ADDR), on 48 bits;
2. The private user key for authentication (link key), on 128 bits;
3. The private user key for encryption, which can be from 8 to 128 bits and is generated by using (among others) the link key;
4. Randomly generated pseudo-random numbers (RAND) on 128 bits.

The BD_ADDR can be obtained via user interactions or automatically when the device is discoverable. The private keys are generated during initialization and are never disclosed. The authentication key is 128 bits long, which means that it is practically impossible to crack, and it lasts as long as the link between devices is established. The encryption key size may vary between 1 and 16 bytes, to allow compliance with different governmental regulations. Even if it is good practice to choose the largest encryption key possible, the encryption key is changed whenever encryption is enforced, which results in a rather short lifetime, hence providing a reasonable protection against key hacking.

The link key can be semi-permanent (i.e., it can be stored in a nonvolatile memory to be used after the current session is terminated) or temporary. Four different types of keys have been defined:

1. The unit key K_A is generated in a single device A, usually hard coded, and is therefore deprecated as it is implicitly insecure, but it will be used for low resource devices and when security is not an issue (e.g.; headsets).
2. The combination key K_{AB} is generated by the two devices but serves the same purpose as the unit key and shall be used whenever a high security level is desired.

3. The temporary master key K_{master} shall only be used during a single session and will replace the original link key temporarily. It will be used, for example, when a master wants to reach more than two devices simultaneously·using the same encryption key.

4. The initialization key K_{init} is used as the link key during the initialization process when noncombination or unit keys have been defined and exchanged, or when a link key has been lost.

Key generation and encryption rely on the SAFER+ algorithm, whose details are given in [1], and follow the scheme described below (also named bonding or pairing):

1. Generation of an initialization key, derived from:
   • The BD_ADDR;
   • The PIN code;
   • A random number of 128 bits.

2. Generation of a (combination) link key:
   • Each unit computes a random number, respectively, LK_K_A and LK_K_B (in fact, a specific combination of a generated random number and its BD_ADDR, to ensure the randomness of the number), and multiplies it (XOR) with the current link key; denote these C_A and C_B.
   • C_A and C_B are exchanged.
   • By multiplying C_A and C_B by the current link key, each device can reconstruct LK_K_A and LK_K_B, and the combination key is generated as $XOR(LK\_K_A, LK\_K_B)$.

3. If no combination key is generated, link keys are exchanged.

4. Authentication is performed by a challenge-response mechanism, in which a claimant's knowledge of a secret key is checked through a two-mode protocol using symmetric secret keys (Figure 3.13). In this scheme, the verifier sends a random number (AU_RAND_A: challenge) to the claimant. The verifier and the claimant perform the same computations by using the secret link key, the random number, and the claimant's Bluetooth address. The claimant sends the result back (SRES), which is verified by the verifier.

5. Generation of the encryption key, based on:
   • The current link key;
   • A random number;
   • A 96-bit ciphering offset number (COF) generated either by the BD_ADDR (when a master key is used) or equal to the ACO, generated during authentication.

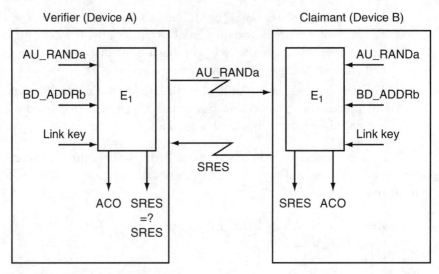

Figure 3.13 Challenge/response scheme.

For point-to-multipoint operation, the master can use separate encryption keys for each slave, although, if the same information has to be sent to two devices, it will have to be sent twice. Hence, for multicast or broadcast operation, the use of a common encryption key (through the use of a master link key) will avoid the overhead of sending the same information more than one time. It should be noted that link keys can be changed if necessary.

Once the key generation and authentication procedures have taken place, encryption can occur, and for this short overview of Bluetooth, it suffices to say that, "there exist good estimates of its strength with respect to presently known methods for cryptanalysis" [1].

3.5.3 Shortcomings and Recommendations

Bluetooth security is mainly based on link layer security, and GAP enables the use of the nonsecure mode, which is probably the main shortcoming of Bluetooth security, albeit inevitable. Recognizing this, the Bluetooth SIG [3] issued two recommendations:

1. Avoid the use of unit keys.
2. Perform bonding in an environment that is as secure as possible against eavesdroppers and use long random Bluetooth passkeys (PINs); indeed, by listening to the exchanges that take part during the pairing process, an attacker could guess the PIN (which also advocates for the use of long PINs).

To ensure good security policies, [3] also describes security architectures for specific profiles, helping the designer to ensure security at the application level. Still, even though Bluetooth specifications enable the design of reasonably secure Bluetooth devices, some attacks (mainly attributable to mistakes in the product design, not to the Bluetooth specifications) have been reported. We will provide some details about these here.

Bluejacking is less a security threat than a little game, that is, if the Bluetooth user is aware of it and is cautious. The technique involves abusing the Bluetooth pairing protocol to pass a message during the initial handshake phase. In this phase, the name (which can be up to 248 bytes long) of the initiator is displayed on the target device; hence, the bluejacker can send some funny messages unnoticed. More seriously, if the pairing goes to the end, the bluejacker can then intrude on the target's device and become a trusted device, possibly having access to the target's data, which can be harmful indeed. The countermeasure is quite simple: be cautious!

Bluesnarfing [5] is the process of "snarfing" (i.e. grabbing information) a Bluetooth-enabled device. This has been wildly publicized on the Internet for mobile phones. By bluesnarfing, an attacker can gain access to important portions of the data stored on the phone, including the phonebook, calendar, business card, International Mobile Equipment Identity (IMEI), and so forth. Even if this is bad news for Bluetooth (or at least for Bluetooth phones), bluesnarfing's flaw is due to a mistake in the implementation of the OBEX profile, where authentication has been omitted, meaning that the flaw can be easily tackled.

The *Bluebug* attack is similar; it is based on the serial profile and enables the use of most AT commands, hence enabling the target's phone for PPP-based networking, sending SMS, and initiating calls.

A simple countermeasure to these attacks is to put the device in undiscoverable mode, so that the attacker is not aware of your presence, even if this somewhat contrary to the WPAN philosophy, where the user wants to have access to his environment in the easiest way possible. Granted, an aggressive attacker could also use *redfang*, a tool that tries to connect to your phone by using all possible BD_ADDRs (the three first bytes of this address are manufacturer specific, and more than half of the Bluetooth enabled phones are Nokia-phones). Of course, this can take a long time, as a pairing process takes about 10 seconds and so is only a threat for nonmobile targets.

The *Backdoor attack* consists of bonding the attacker's device to the target in a first phase (necessitating the target's consent) and then making sure that the pairing information is not visible to the target's user. Hence, the attacker can then have access to most of the resources of the target, as it is considered trusted. Although this is mostly an implementation flaw of the specifications, these specifications could introduce more constraints on the lifetime of the pairing and hence somewhat alleviate this problem.

Denial of service attacks can be performed on Bluetooth by flooding data or by repeated pairing attempts. Still, up till now the only serious problem reported (but not really a security threat) is that some forged Bluetooth packets have caused some phones to crash and reset, without doing any more harm.

Bluetooth tracking is not a security threat in itself, but more a problem of protecting one's private life. Indeed, once you have discovered a Bluetooth-enabled device, with appropriate material (e.g., a good antenna and another Bluetooth-enabled device), you can track somebody's displacements, violating his private life.

3.5.4 Conclusions on Security

Bluetooth specifications offer good security at the link level but still allow some major flaws to appear in Bluetooth-enabled mobile phones, due to some mistakes in the design. Still, by the very nature of Bluetooth, devices are exposed to many potential attackers, which advocates for the following:

- Users should be made security aware. Most users still have a 0000 PIN code and are not aware that this can open their phone to abuse.

- Application-oriented security should be used when security is an issue.

3.6 Future of Bluetooth: Team Up with UWB

Bluetooth has evolved through several cycles of hope and disillusionment. Technically it has grown to a rather complex and complete solution and can be said to have reached its goal: that is, to provide an easy solution for cable replacement at an affordable cost (less than $5 per device). As such, its industrial near future is secured. The long-awaited Bluetooth 2.0 arrived late 2004, but it is only a marginal improvement over Bluetooth 1.2, at least in terms of rate (from 1 to 3 Mbps) and capabilities.

From the point of view of its evolution, Bluetooth will have to face serious competitors like 802.15 (presented in Chapter 4), both in the very low and very high data rate applications. If Bluetooth SIG wants to broaden its scope in the multimedia area, it will have to compete or team up with 802.15 WPAN standards. In early May 2005, Bluetooth SIG announced its intent to work with the developers of ultra-wideband (see Chapters 4 and 6 for details) to combine the strengths of both technologies. According to the SIG, "this decision will allow Bluetooth technology to extend its long-term roadmap to meet the high-speed demands of synchronizing and transferring large amounts of data as well as enabling high quality video applications for portable devices. UWB will benefit from Bluetooth technology's manifested maturity, qualification program, brand equity and comprehensive application layer."

The common work between UWB and Bluetooth is good news for WPANs, since it means that the ease of use of Bluetooth, thanks to its large collection of profiles, will be complemented with very high data rates to offer easy-to-use multimedia WPAN systems and devices.

References

[1]   Bluetooth Specification Including Core v2.0, November 2004, available at http://www.bluetooth.org.

[2]   "Enhanced Rate Addendum to the Core Specifications," April 27, 2004, available at http://www.bluetooth.org.

[3]   Gehrmann, C., "Bluetooth Security White Paper," Bluetooth SIG Security Expert Group, April 19, 2002, available at http://www.bluetooth.org.

[4]   Withehouse, O., "War Nibbling: Bluetooth Insecurity at Stake," research report, October 2003.

[5]   Laurie, A., "Serious Flaws in Bluetooth Security Lead to Discolsure of Personal Data," A.L. Digital Ltd., November 2003, available at http://www.bluesteumbler.org.

[6]   Niem, Tu, C.,"Bluetooth and Its Inherent Security Issues," SANS Giac Security Essentials Certification v1.4b, April 2002.

[7]   ETSI Technical Specification 07.10, available at http://webapp.etsi.org/pda/home .asp?wki_id=9198.

4

802.15 Overview

4.1 Introduction

At the end of the 1990s, following the expected success of Bluetooth, IEEE and the industry recognized the need for WPANs and began a standardization effort. Initially, there were four task groups, leading to four standards:

1. 802.15.1 [1] provides an IEEE version of Bluetooth 1.2 and 2.0.

2. 802.15.2 [1] provides rules for coexistence between 802.15 families and other wireless standards (mostly 802.11) to avoid (or manage) interferences.

3. 802.15.3 [2–4] provides a standard for high rate (20 Mbps or more) WPANs. The task group defined the PHY and MAC specifications for high data rate wireless connectivity with fixed, portable, and moving devices within or entering a personal operating space (POS). The main features and characteristics of the draft standard are as follows:
   • Data rates: 11, 22, 33, 44, and 55 Mbps;
   • QoS isochronous protocol;
   • Ad hoc peer-to-peer networking;
   • Security;
   • Low power consumption;
   • Low cost;
   • Designed to meet the demanding requirements of portable consumer imaging and multimedia applications.

4. 802.15.4 [5–8] specifies the medium access control and the physical layer for low rate WPANs. A low rate WPAN is a simple, low-cost communication network that allows wireless connectivity in applications

with limited power and relaxed throughput requirements. The main objectives of a low rate WPAN are ease of installation, reliable data transfer, short-range operation, extremely low cost, and a reasonable battery life, while maintaining a simple and flexible protocol. Some of the most important characteristics of a low rate WPAN specified in the IEEE 802.15.4 standard are:

- Over-air data rates of 250, 40, and 20 Kbps;
- Star or peer-to-peer operation;
- Allocation of guaranteed time slots;
- CSMA-CA channel access (in a classical CSMA scheme, the device listens to the channel and transmits its data as soon as the channel is idle; in the CSMA-CA scheme, once the channel is idle, the device waits for a random back-off time and starts to transmit its data—if the channel is still idle);
- Fully acknowledged protocol for transfer reliability;
- Low power consumption;
- Energy detection;
- Link quality indication;
- 16 channels in the 2,450-MHz band, 10 channels in the 915-MHz band, and 1 channel in the 868-MHz band.

There are additionally five task/study groups working on alternate PHY/MAC proposals for both high rate and low rate WPANs and on mesh networking:

1. The IEEE 802.15 high rate alternative PHY task group (TG3a) is working to define a project to provide broadband WPAN up to 480 MHz, based on UWB technology. Currently, two main proposals are fighting fiercely against each other, one based on multiband OFDM technology and one based on direct sequence UWB, and an attempt is being made to merge them into one standard, where the two PHY layers could coexist.

2. The IEEE 802.15 low rate alternative PHY task group (TG4a) is working on very low rate WPANs, also based on UWB technology.

3. The IEEE 802.15 low and high rate maintenance groups (TG3b and TG4b) are working on optimizations of the current standards.

4. The IEEE 802.15 mesh networking task group (TG5) will determine the necessary mechanisms that must be present in the PHY and MAC layers of WPANs to enable mesh networking and to provide the following characteristics:
   - Extension of network coverage without increasing transmit power or receive sensitivity;

- Enhanced reliability via route redundancy;
- Easier network configuration;
- Better device battery life due to fewer retransmissions.

5. The IEEE 802.15 millimeter wave study group (SG3c) is developing a millimeter wave–based alternative physical layer for the high rate WPAN. This millimeter-wave WPAN will operate in the 57- to 64-GHz unlicensed band. The 60-GHz WPAN will allow high coexistence (close physical spacing) with all other microwave systems in the 802.15 family of WPANs, as well as very high data rates in excess of 2 Gbps.

The rest of this chapter will detail the 802.15.3 and 802.15.4 standards—as Bluetooth is described elsewhere and the details of coexistence are beyond the scope of this book. We will also touch upon UWB, which is at the heart of TG3a's battle and is also the candidate for 802.15.4a, the very low data rate alternative to 802.15.4 (TG4a).

4.2 802.15.3: The High Rate WPAN

The high rate WPAN (HR-WPAN) is designed to support ad hoc networking, multimedia QoS provisions, and power management (unlike the original 802.11 WLANs, with the exceptions of 802.11e, which provides support for QoS, enabling voice over Wi-Fi, and 802.11f, which provides some support for power management and interoperability). The basic building block is the piconet, and an HR-WPAN can be formed by one independent piconet, by one parent piconet, and a collection of child piconets. The higher networking layers are not defined in the 802.15.3 specifications; they are determined by the user's definition (or other industrial standards). Routing support for mesh networks will be defined by Task Group 5. The MAC layer provides support for QoS by using a mix of CSMA/CA (for the best effort traffic) and TDMA (for guaranteed data rate support). The PHY layer operates in the 2.4-GHz band and achieves data rates from 11 to 55 Mbps, based on classic trellis coded PSK/QAM modulation.

4.2.1 The High Rate Network

The HR-WPAN is based on piconet architecture. The piconet is defined as a wireless ad hoc data communications system that allows a number of independent data devices (DEVs) to communicate with each other. The piconet operates in a small area around a person or object that typically covers at least 10m in all directions, enveloping the person or object whether stationary or in motion.

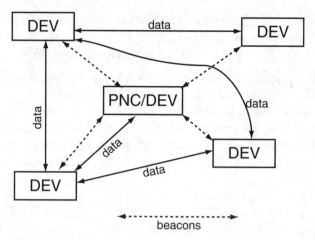

Figure 4.1 A typical 802.15.3 piconet.

A 802.15.3 piconet consists of several components, as shown in Figure 4.1.

The basic component of a piconet is the DEV. One DEV is required to assume the role of the piconet coordinator (PNC). The PNC provides the basic timing for the piconet with the beacon. Additionally, the PNC manages QoS requirements, power save modes, and access control to the piconet. This standard allows a DEV to request the formation of a subsidiary piconet. The original piconet is referred to as the parent piconet. The subsidiary piconet is referred to as either a child or neighbor piconet, depending on the method the DEV uses to associate with the parent PNC. Child and neighbor piconets are also referred to as dependent piconets since they rely on the parent PNC to allocate channel time for the operation of the dependent piconet. The difference between a child and neighbor piconet is that communication is possible between devices of child and parent piconets, but not between a neighbor and parent. An independent piconet is a piconet that does not have any dependent piconets.

4.2.2 Overview of MAC

IEEE 803.15.3 MAC is designed to support the following goals:

- Fast connection time;
- Ad hoc networks;
- Data transport with QoS;
- Security;
- Dynamic membership;
- Efficient data transfer.

Before describing the operation and management of a piconet, we will detail the superframe structure, which will show us how QoS is managed in 802.15.3.

4.2.2.1 Superframe Structure

802.15.3's MAC operation is based on a PNC and DEVs. All communications between these entities take place in a superframe (Figure 4.2), which is divided into three main parts:

1. The *beacon*, which is sent only by the PNC to other devices (or to a child/neighbor PNC), is used to set the timing allocations and to communicate management information for the piconet. The beacon consists of the beacon frame, as well as any announce commands sent by the PNC. Hence, the 802.15.3 beacon is similar to the broadcast channel used in cellular networks.

2. The *contention access period* (CAP) is a period in which all DEVs can contend for the channel, using a CSMA/CA scheme. This is similar to 802.11 operation mode, except that:
   - The data communications are peer to peer (two DEVs can exchange data directly, without having to transfer the data through the PNC);
   - There is no optional request to send/clear to send (RTS/CTS) exchange, which means that the network can be subject to hidden terminal and exposed terminal problems.

   Note that management frames can also be transmitted in this CAP.

3. The *channel time allocation period* (CTAP) operates in a TDMA fashion. The CTAP is divided into several CTAs (similar to the guaranteed time slots in 802.15.4), which can carry commands and asynchronous data, but are meant to carry isochronous data transfers. Indeed, a device desiring to transmit data with a certain QoS (e.g., a fixed data rate) asks the PNC for a given data rate, and the PNC can grant it by allocating a sufficient number of CTAs in each superframe.

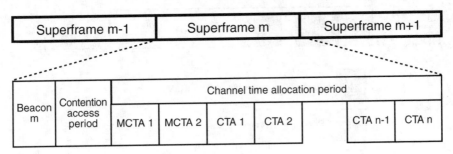

Figure 4.2 The 802.15.3 superframe structure.

4.2.2.2   Death and Birth of a Piconet (Family)

A piconet is formed when an 802.15.3 DEV that is capable of acting as the PNC begins transmitting beacons. The process is rather simple. The candidate PNC scans its environment, and then:

1. If there is no 802.15.3 piconet around, it starts the piconet by sending beacons;
2. If there is an existing piconet, it can:
   - Join the piconet. In this case, if it has more capabilities than the current PNC, and if the security policy allows it, the PNC candidate can become the new PNC by the use of a handover process, which maintains all existing time allocations to ensure soft transition from one PNC to the other.
   - Become a child piconet (the existing piconet then becomes the parent piconet). The child piconet uses a distinct piconet ID and acts as an autonomous piconet but depends on a private CTA from the parent piconet, as shown in Figure 4.3. The principle is rather simple: during the CTA period, the parent piconet first negotiates the resources with the child piconet (during a first slot) and considers the communication period of the child piconet as a private CTA. The child piconet can use the allocated time in a classical way (hence, the child piconet management, especially piconet membership, is independent of the parent piconet). During the parent piconet's normal operation, the child piconet considers the time to be reserved, so that there is no interference between parent and child piconets. Communication can take place between child and parent devices.

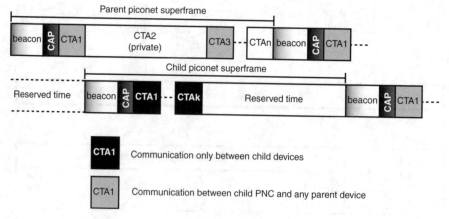

Figure 4.3   Relation between the parent and child piconet superframes.

- Become a neighbor piconet. A neighbor piconet behaves essentially in the same way as a child piconet, with the difference that no communication can take place between devices of a parent piconet and its neighbor piconet.

When a PNC wants to leave its piconet (e.g., for shut-down), a handover process is started to elect another DEV as the new PNC. If there is no PNC-capable device in the piconet, the piconet is ended. Note that procedures are defined to handle the case of dependent (i.e., child or neighbor) piconets as well as the most common incidents (such as power loss).

4.2.2.3 Data Transfers and Service Discovery

Data transfers can be made robust by the use of sequence numbers, variable size fragmentation, and by the following acknowledgments:

- No-ACK policy, in which no guarantee is offered;
- Immediate-ACK policy, in which each frame is individually and immediately acknowledged;
- Delayed-ACK policy, in which the source requests explicitly an acknowledgment for a group of frames. The goal is to decrease the overhead caused by the acknowledgments. Note that delayed-ACK should be used for isochronous communications (e.g., video/audio), and that for real-time isochronous communications, the source can decide not to retransmit the unacknowledged frames.

Since 802.15.3 piconets are ad hoc in nature, it is important for the DEVs in the piconet to be able to find out information about the services and capabilities of the other DEVs in the piconet at any instant in time.

Discovery methods are based on requests and announcements. Basically, a DEV can find information on the whole piconet or on a specific device by requesting this information to the PNC or requesting specific information from a DEV. A DEV can also announce its capabilities to any other DEV or the PNC.

Additionally, to allow better quality management (e.g., power management, interference mitigation, and channel selection), the PNC can request channel state information (how good is a specific channel) either by direct measurement or by asking the DEVs attached to the piconet.

4.2.2.4 Interference Mitigation and Power Management

The piconet operates in a dynamic environment and under unlicensed operation rules. Hence, it is subject to interference from other 802.15.3 piconets as well as other unlicensed wireless entities using its channels. With the benefits of the channel state information it gathers, the PNC can dynamically change the channel used by the piconet and move it, if needed, to a new channel without disrupting the piconet's operation.

Moreover, to minimize interference with other wireless networks that share the same channel (and to decrease the power consumption), the piconet can use power control. Indeed, when two devices are close to each other, they need a small amount of power to receive data correctly. Using the minimum amount of power, which is the essence of power control, has two major advantages: less interference is generated for the other users and less power will be consumed.

802.15.3 devices should be able to change their power by increments of 3 dB, as soon as the transmitted power is higher than +4 dBm (until the maximum power, which is 20 dBm, that is, 100 mW). The power control can take place both in a centralized way (defined by the PNC) or in a decentralized way (two communicating devices in a CTA decide to decrease their power).

Finally, to minimize power usage for battery-powered devices, three management techniques are used:

- Device synchronized power save (DSPS) mode, in which groups of DEVs are allowed to sleep for multiple superframes. Based on the DSPS information, devices can wake up together to transfer data between members of the group, or wake up when another group needs to send data to one of its members.;

- Piconet-synchronized power save (PSPS) mode, in which DEVs are allowed to sleep at intervals defined by the PNC. In this mode, the devices have to listen for wake beacons at specified times to be awakened if needed.

- Asynchronous power save (APS) mode. In this mode, a device has to communicate with the PNC before a certain maximum limit to preserve its membership, but may sleep otherwise.

Regardless of the DEV's power management mode, every DEV in the piconet is allowed to power down during parts of the superframe when the DEV is not scheduled to transmit or receive data.

4.2.2.5 Security Overview

A piconet can be in one of the two following security modes:

- *Mode 0—Open:* Security membership is not required and payload protection (either data integrity or data encryption) is not used by the MAC. The PNC is allowed to use a list of DEV addresses to admit or deny entry to the piconet.

- *Mode 1—Secure membership and payload protection:* Devices must establish secure membership with the PNC before they have access to the piconet's resources. Data sent in the piconet is allowed to use payload protection (data integrity and/or data encryption). Data integrity is required for most of the commands that are sent in the piconet. Encryption is based on AES-128 symmetric keys. Additionally, there are potentially four different keys used in a piconet:

  1. The PNC-DEV management key, used for most commands between a PNC and a secure member (there are as many PNC-DEV keys as there are secure DEVs);

  2. The piconet group key, used for all data communications as well as for some management frames (e.g., broadcast information frames);

  3. The peer-to-peer management key, used for most commands between two devices that have established a peer-to-peer secured communication. This allows protection from other members of the piconet;

  4. The peer-to-peer data key, used for data encryption between two devices that have established a peer-to-peer secured communication.

4.2.3 The Physical Layer

802.15.3 uses the unlicensed 2.4- to 2.4835-GHz band, which is divided into five bands, each 15 MHz wide. In high-density networks, four of these five bands can be used; otherwise, for better coexistence with 802.11b networks, three bands should be used (see Table 4.1).

In a channel, data is sent at a symbol rate of 11 Msps (11 Mbauds). The basic data rate is 22 Mbps, using DQPSK modulation and no coding. Optionally (but probably in most devices), 802.15.3 devices may use additional rates up to 55 Mbps, using trellis coded modulation [9] and M-QAM modulations, as defined in Table 4.2 and Figure 4.4. TCM uses a higher order modulation

Table 4.1
802.15.3 Frequency Plan

Channel ID	Center Frequency	High Density	802.11b Coexistence
1	2.412 GHz	X	X
2	2.428 GHz	X	—
3	2.437 GHz	—	X
4	2.445 GHz	X	—
5	2.462 GHz	X	X

Table 4.2
802.15.3 Rates and Modulations

Modulation Type	Coding	Data Rate
QPSK	8-state TCM	11 Mbps
DQPSK	No	22 Mbps
16-QAM	8-state TCM	33 Mbps
32-QAM	8-state TCM	44 Mbps
64-QAM	8-state TCM	55 Mbps

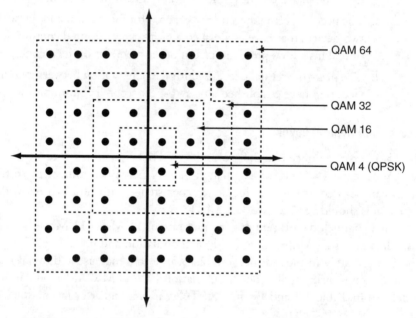

Figure 4.4 802.15.3 constellations.

combined with trellis coding, resulting in better performance (in terms of bit error rate). For more details on trellis coded modulation, refer to [9].

The PHY frame is classic, as indicated in Figure 4.5. It consists of the preamble, using a constant-amplitude, zero-autocorrelation (CAZAC) sequence [10] and the header, which is protected by a correcting code; these are sent at the default 22 Mbps. The payload itself is sent at the nominal rate (from 11 to 55 Mbps).

On the receiver side, the sensitivities, which are similar to what is required for 802.11b, are given by Table 4.3.

Normal operation

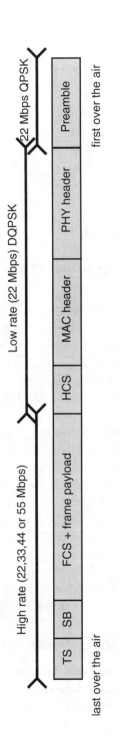

Noisy environment operation

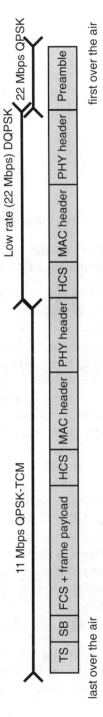

Figure 4.5 802.13.3 PHY frame.

Table 4.3
802.15.3 Sensitivities

Rate	Sensitivity
11 Mbps	−82 dBm
22 Mbps	−75 dBm
33 Mbps	−74 dBm
44 Mbps	−71 dBm
55 Mbps	−68 dBm

4.3　TG3A: The (Very) High Rate WPAN

4.3.1　Ultra-Wideband: An Old Technology for High Rate Communications [11, 12, 17]

Only a few years ago, hardly anybody would have dreamed of sending 1 Gbps over the air with consumer equipment. Today, some proposals in TG3a go beyond that number, and this dream is rapidly becoming true, thanks to the oldest wireless technology. Indeed, the origin of UWB wireless systems trace back to Hertz's 1893 experiment, which used a spark discharge to generate electromagnetic waves. Marconi used this device to realize the first wireless link in 1897, in Genoa, Italy. By using sparks, which translate in short electromagnetic pulses, the communication link uses a large bandwidth (roughly inversely proportional to the pulse's duration), unlike classical narrowband communications.

At the beginning of the 20th century, in order to be able to share the spectrum between different communications, engineers decided to use a collection of narrow bands, each band carrying a link. This paradigm has been prevalent until now, with some variants (notably spread spectrum). Still, during World War II, impulse radio (IR) systems (i.e., systems based on the emission of short pulses) were developed, mainly for radar applications, as the small size of the pulses also translates to good localization capabilities. In the 1990s, with the progress of microelectronics and with a better understanding of system characteristics, low-cost IR-based wireless communication systems were built and commercialized.

4.3.2　Ultra-Wideband: When Shannon Meets Marconi (or Was it Fourier?)

In the quest for large data rates, IR, or more exactly the use of a very large bandwidth, promises to be the right solution, as can be seen from Shannon's capacity formula. Indeed, the capacity of a channel can be written as

$$C = W \cdot \log_2(1 + SNR) \tag{4.1}$$

where C is the capacity in bits per second, W is the bandwidth in Hertz, and SNR is the signal to noise ratio (i.e., the ratio between the power of the signal and the power of the noise). Hence, the capacity increases *logarithmically* with the power (as the noise power remains constant in a given bandwidth) but *linearly* with the bandwidth. Thus, Shannon tells us that it is more efficient to use a larger bandwidth with a small power than to use a large power in a small bandwidth. This is illustrated by Figure 4.6, where the capacity is plotted against the bandwidth. This figure shows that, even for very low power spectral density (−41.3 dBm/MHz, which is less than 0.1 mW for 1 GHz), the capacity of UWB exceeds that of popular WLAN systems as soon as more than 1-GHz bandwidth is used.

Note that the Shannon relation only gives us the limit that cannot be surpassed; it does not tell which modulation/coding has to be used to reach this limit. Initially, UWB was introduced as IR (sending very short pulses), but other schemes based on orthogonal frequency division modulation (OFDM; used, for example, in 802.11a and g, as well as in ADSL) have been proposed.

4.3.3 Ultra-Wideband: From Regulatory Acceptance to Standards

Besides the capacity advantage described earlier, the major advantage of UWB is that, due to the use of very low spectral density, it can be used on top of other

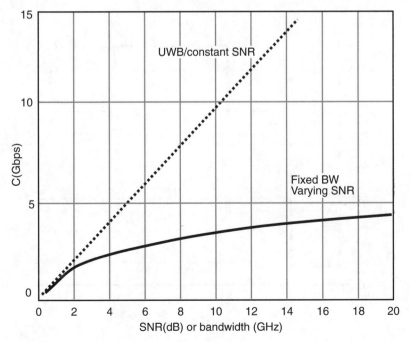

Figure 4.6 Capacity of an UWB channel with a power spectral density of −41.3 dBm/MHz.

narrowband communications, hence using the spectrum twice! To allow the use of such a scheme, the regulators had to ensure that UWB would not cause harmful interference to other users [the dispute was particularly important for global positioning systems (GPS), which also use low power densities]. After several years of negotiations, the FCC on February 14, 2002, permitted the marketing and operation of UWB, defining it according to the following criteria:

1. The fractional bandwidth must be larger than 20% (25% in the initial text) or the bandwidth be larger than 500 MHz (1.5 GHz in the initial text). Note that the fractional bandwidth is defined as $2(f_H - f_L)/(f_H + f_L)$ where f_H is the upper frequency of the –10-dB emission point and f_L is the lower frequency of the –10-dB emission point. The center frequency of the transmission was defined as the average of the upper and lower –10-dB points [i.e., $(f_H + f_L)/2$].

2. UWB must use a very low power spectral density, allowing, in the band used for communication, an average power less than –41.25 dBm/MHz (FCC Part 15 unintentional emission limit) and a peak power less than 0 dBm in a 50-Mhz bandwidth.

3. Approved spectrum is application specific. For WPAN systems, this boils down to the band below 960 MHz and the band between 3.1 and 10.6 GHz, hence offering 7.5-GHz bandwidth (Figure 4.7).

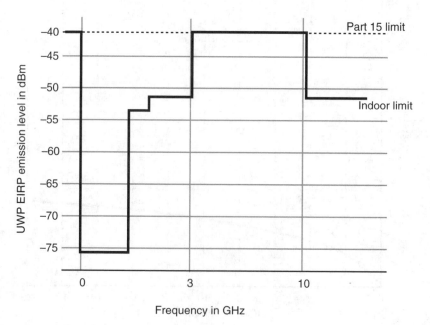

Figure 4.7 UWB spectrum limits for indoor operation.

With the FCC's approval, TG3a could really start its work on defining a new standard, with the goal to offer 110 Mbps on a distance of 10m, 200 Mbps on a distance of 4m, and up to 480 Mbps on shorter distances. After the usual selection process, two main proposals still remain, one based on impulse radio (the DS-UWB proposal) and the other based on OFDM (the MB-OFDM proposal).

4.3.4 The DS-UWB Proposal

The historical origins of UWB lie in sending short impulses [12], which is also the technique used by DS-UWB. In DS-UWB, signals are typically modulated pulse trains, with very short pulse duration and uniform or nonuniform interpulse spacing.

The essence of UWB, and the differences with classical narrowband, is illustrated in Figure 4.8. The top of the figure shows one chip of a CDMA system, with a carrier frequency of 1.8 GHz and a bandwidth of about 3.84 MHz, leading to a fractional bandwidth of 0.019%. The following line of the figure shows a (hypothetical) signal waveform, whose fractional bandwidth is about 6%, which is still considered narrowband. From 20% fractional bandwidth on, the signal is considered as ultra-wideband, the last line showing the traditional ultra-wideband signal, made of a (usually Gaussian) single pulse, which translates in a bandwidth spanning the whole spectrum, up to the inverse of the pulse duration (here, starting from dc, where the spectrum is null thanks to the symmetry of the pulse, up to 10 GHz for a 100-ps-wide pulse).

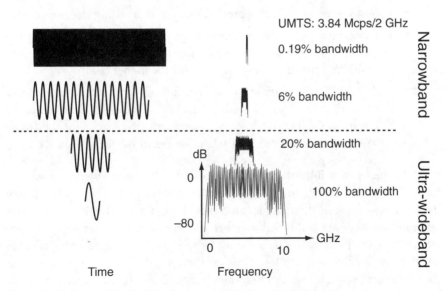

Figure 4.8 Pulses and bandwidths from narrowband to ultra-wideband.

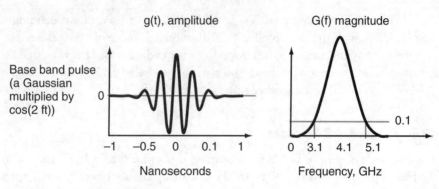

Figure 4.9 A passband UWB signal.

When comparing the classical UWB spectrum with the spectrum allocated by the FCC, it is obvious that some work has to be done on the original UWB signal to comply with the regulations. A simple way to do this is to shift the signal to the desired spectrum by multiplication by a simple sinusoid, like that sketched in Figure 4.9. The main advantage, compared to classical spectrum shifting used in narrowband communications, is that this operation does not need to be performed by an expensive mixer, but rather by generating directly the pulse needed with a template. In the DS-UWB proposal, two bands are proposed, one from 3.1 to 5.15 GHz and one from 5.825 to 10.6 GHz. The main advantages of this spectral mode of operations are as follows:

- It avoids the 5.15- to 5.825-GHz band, where 802.11a lies.
- It allows good neighboring piconet separation by using the two bands.
- It allows the use of two bands targeted at different usage [low rate (from 28 to 400 Mbps] long-range services for the lower band; high rate (from 57 to 800 Mbps) short-range services for the higher band].
- It allows frequency duplex operation (sending, for example, data in one direction in the 3.1- to 5.15-GHz band and in the other direction in the 5.825- to 10.6-GHz band, by using different frequencies for the cosine multiplied base band pulse illustrated in Figure 4.9).

Another major advantage of pulse-based UWB is the absence of what is called Raleigh fading (Figure 4.10). Indeed, if a device receives two replicas of the same signal, and if these signals are very close in time compared to the symbol duration, they are combined together and interfere at the receiver. This interference gives birth to Raleigh fading, where the signal is received with an apparent power that is much lower than if there was only a direct path. Due to the very short duration of the pulses, this effect is almost totally absent in UWB, hence allowing a lower emitting power.

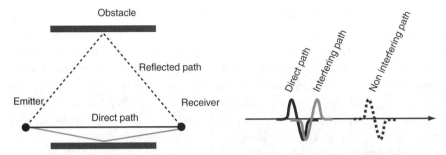

Figure 4.10 Path 2 and the direct path interfere, but due to the shortness of the pulses, path 1 and the direct path do not interfere. Globally, the amount of interference in UWB is much lower than for narrowband communications.

To carry the data on this type of pulse, DS-UWB uses simple BPSK and length 24 (for the lowest rates) and 12 (for the highest rates). The combination of these modulations and various spreading and coding schemes leads to bit rates from 28 Mbps (in the lower band) to 1,320 Mbps (using the two bands). Moreover, by using the two different bands and treating the spreading codes like channelization codes, 16 simultaneous piconets can be created (as each piconet relies on a different channel).

4.3.5 The MB-OFDM Proposal

The multiband OFDM proposal is based on OFDM modulation. OFDM is now a 40-year-old multicarrier technology [13, 14], and it is present in the popular ADSL as well as in high rate WLANs (802.11g and 802.11a) and other wireless standards. The principle of OFDM is based on sending multiple data on multiple carriers, as depicted in Figure 4.11.

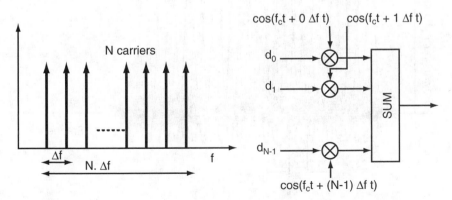

Figure 4.11 The principle of multicarrier modulation.

The main advantage of multicarrier modulation is that the data symbols transmitted have a large duration T, compared to their single carrier counterpart (indeed, the duration T for the multicarrier symbol is N_c times the single carrier symbol duration); hence, the time spreading due to the channel produces less intersymbol interference (ISI). Moreover, to yield optimal spectral efficiency, the carriers are chosen to be orthogonal, allowing the individual bands to overlap, whereas the receiver can perfectly separate them thanks to their orthogonality (see Figure 4.12 for a sketch of the spectral representation).

OFDM was really implemented when it was recognized that the modulator (and demodulator) could be implemented by the use of a fast Fourier transform (FFT), leading to a low complexity implementation.

The MB-OFDM relies on this low complexity and proposes the use of a 128-carrier OFDM signal, spanning a bandwidth of 528 MHz. The MB-OFDM offers data rates ranging from 53.5 to 480 Mbps, thanks to various coding and modulations on the carriers. Moreover, to span (eventually) the whole spectrum offered by the FCC, it proposes a multiband approach, where the signal will hop from one frequency to the other according to a specified sequence. In addition, similarly to Bluetooth, different hopping sequences offer multiple channels, allowing the collocation of 16 different piconets. (See Figure 4.13.)

Without going into the details of the dispute between DS-UWB and MB-OFDM, suffice it to say that MB-OFDM proponents argue that OFDM is a proven technology that enables, by putting some carriers to zero, the precise sculpting of the spectrum, which is useful for interference avoidance.

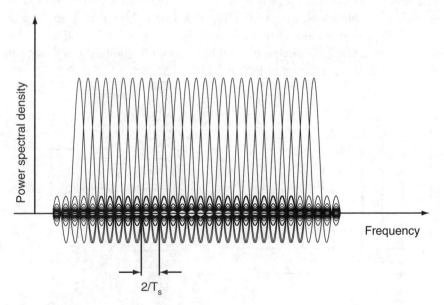

Figure 4.12 Spectral representation of an OFDM signal.

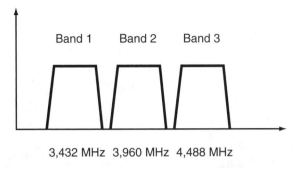

Figure 4.13 MB-OFDM frequency of operation for mode 1 devices.

4.3.6 TG3a Conclusion

TG3a defenders propose a two-PHY-based merged proposal, allowing both PHY layers (and industry groups) to find their way through the standards and ensuring a minimal interoperability mode as well as a common MAC layer. This will lead to a common standard with two different modes, each optional, and a common signaling system that enables devices based on two different PHY layers to exchange data at low rate (about 10 Mbps).

4.4 802.15.4: The Low Rate WPAN

802.15.4 is targeted at low rate wireless personal area networks (LR-WPANs) and is aimed at providing ultra-low complexity, ultra-low cost, and ultra-low power consumption devices. It supports over the air data rates of 20, 40, and 250 Kbps, providing enough data rate for simple interactive toys or low-quality multimedia, down to sensor applications. On the physical layer, 802.15.4 uses direct sequence spread spectrum in the 868-Mhz (1 channel/20 Kbps) to 915-MHz (10 channels/40 Kbps) band as well as in the 2,450-MHz band (16 channels/250 Kbps). The physical layer provides energy detection (ED) and link quality indication (LQI).

On the MAC layer, 802.15.4 uses a uses CSMA/CA medium access mechanism. Transfer reliability is ensured by a fully acknowledged protocol. Devices are addressed by 16-bit short addresses inside PANs (allowing for PANs of up to 65,536 devices) or by unique 64-bit extended addresses (similar to classical MAC addresses, and allowing for networks of up to 2^{64} devices). So with up to 65,536 locally addressed devices, 802.15.4 is well suited for large and dense sensor networks.

An optional superframe structure provides some QoS support for real-time operation, the PAN coordinator being able to provide guaranteed slots (GTSs).

The network topologies are the star and the peer-to-peer topologies. To allow for very low-complexity/low-power devices, two different device types are defined: the full-function device (FFD) and the reduced-function device (RFD). The FFD can be used as a PAN coordinator or as a simple device. An RFD can only talk to an FFD (in a star topology), but FFDs can also talk to other FFDs (in a peer-to-peer topology).

Basic security mechanisms are foreseen and supported at the link layer.

4.4.1   Network Topologies

The LR-WPAN may operate in either of two topologies: the star topology or the peer-to-peer topology (Figure 4.14).

In both topologies, the communication is controlled by the PAN coordinator, which is responsible for the creation of the network and assignment of the unique PAN identifier, has routing duties, and possibly runs some WPAN-specific application. The PAN coordinator can also allocate short addresses to the devices. In a star topology, all communications have to go through the PAN coordinator, unlike in the peer-to-peer network, which allow for mesh networks and thus more complex ad hoc networks, including multiple hop networks (not included in the 802.15 standards to date).

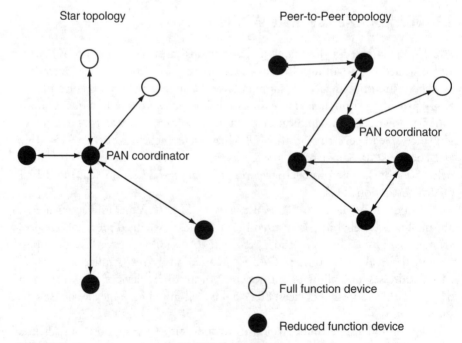

Figure 4.14 Star and peer-to-peer LR-WPAN topologies.

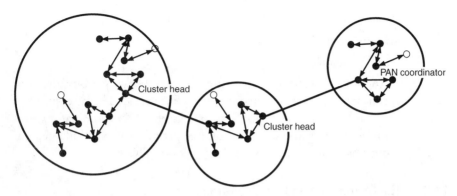

Figure 4.15 Cluster tree network.

Network formation is performed by the network layer. Like for Bluetooth (but for which the star formation takes place in the link layer), the star network is formed by the first FFD waking up and taking the PAN coordinator role, thus choosing a PAN identifier and allowing other devices to join the network. In a peer-to-peer network, one can build complex network topologies comprising different clusters in which other FFDs take a coordination function (not PAN coordinator, but *simple* coordinator) and act as a cluster head (Figure 4.15). The obvious advantage of this topology is the increase in coverage, while the disadvantages are the reduced network capacity [15] and the increased latency.

4.4.2 The MAC Sublayer

Like in all 802 standards, the data link layer (DLL) is split into two sublayers, the MAC and the logical link control (LLC) sublayers. The LLC sublayer is standardized in 802.2. The MAC layer itself is specific and takes care of beacon management, channel access, GTS management, frame validation, acknowledged frame delivery, association, and disassociation. In addition, the MAC sublayer provides hooks for implementing application-appropriate security mechanisms. One of the striking characteristics of 802.15.4 is its simplicity (less than 30 primitives), which allows a very low-cost implementation (holds on less than 32 KB of memory).

The LR-WPAN MAC is based on the following:

- A coordinator;
- Four types of MAC frames:
1. A beacon frame, which bounds the superframe (all networks are not based on superframes, and so there can be beaconless networks, which help to lower the power consumption);

2. Data frames (maximum 127 bytes, including headers);

3. Acknowledgment frames;

4. MAC command frames.

- An (optional) superframe structure, divided in two periods: the contention access period and the contention free period. The latter provides a mechanism to guarantee QoS, as follows: the PAN coordinator grants a number of slots in the contention free period to a given link, guaranteeing a data rate, as these slots arrive periodically.

4.4.2.1   The Coordinator and the Superframe

As indicated here above, the coordinator is responsible for the creation and the management of the network. In a star network, all communication transits through the coordinator; in a mesh network, the coordinator assigns time slots to devices that can then communicate directly together. This leads to three types of data transfer transactions: transfer from a coordinator, transfer to a coordinator, and transfer between two peer devices.

For beacon-enabled networks, the coordinator defines the superframe structure. The superframe is bounded by network beacons, sent by the coordinator (see Figure 4.16), and is divided into 16 equally sized slots. The beacon frame is transmitted in the first slot of each superframe. The beacons are used to synchronize the attached devices, to identify the PAN, and to describe the structure of the superframes. Any device wishing to communicate during the contention access period between two beacons shall compete with other devices using a slotted CSMA/CA mechanism.

The coordinator may dedicate portions of the superframe to devices requiring guaranteed data rate or delay response, in the form of up to a total of seven GTSs per superframe, in the contention free period. For low-latency applications or applications requiring specific data bandwidth, the PAN coordinator may dedicate portions of the active superframe to that application.

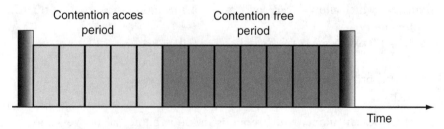

Figure 4.16 The superframe structure may comprise a contention free period, providing some QoS.

4.4.2.2 The Frame Structure

The frame structures have been designed to keep the complexity to a minimum, while at the same time making them sufficiently robust for transmission on a noisy channel. The LR-WPAN defines four frame structures:

1. A beacon frame, used by a coordinator to transmit beacons;
2. A data frame, used for all transfers of data;
3. An acknowledgment frame, used for confirming successful frame reception;
4. A MAC command frame, used for handling all MAC peer entity control transfers.

The generic structure of the frames is given in Figure 4.17.

4.4.2.3 Multiple Access and Acknowledgment

The beaconless networks use an unslotted CSMA/CA scheme. In a classical CSMA scheme, the device listens to the channel and transmits its data as soon as the channel is idle. In the CSMA/CA scheme, once the channel is idle, the device waits for a random back-off time and starts to transmit its data (if the channel is still idle).

The beacon-enabled networks, in the contention access period, use a slotted CSMA/CA scheme, where the frame fits in a time slot and has to start at predefined instants following the beacon.

In the contention free period, the coordinator broadcasts the assigned GTSs in the beacons, and the transmissions take place in these slots without contention. To get these GTSs, a device must first negotiate with the coordinator (during the contention access period).

To ensure that a transmission was successful, an optional acknowledgment mechanism is foreseen. To receive an acknowledgment (confirming that a data frame or a MAC command frame was successfully received), the originating device requests it by setting the corresponding bit in the frame header. If no ACK is received, the originating device may resend its packet until proper reception.

4.4.2.4 Security

As in most wireless standards, basic security services are provided at the MAC level. Security is based on the (optional) use of symmetric keys, with freshness protection (making sure that keys are changed often enough, this service being provided by higher layers) and management of an ACL. The main security features are as follows:

- *Access control:* The access control list is the list of devices from which a specific device expects to receive frames.

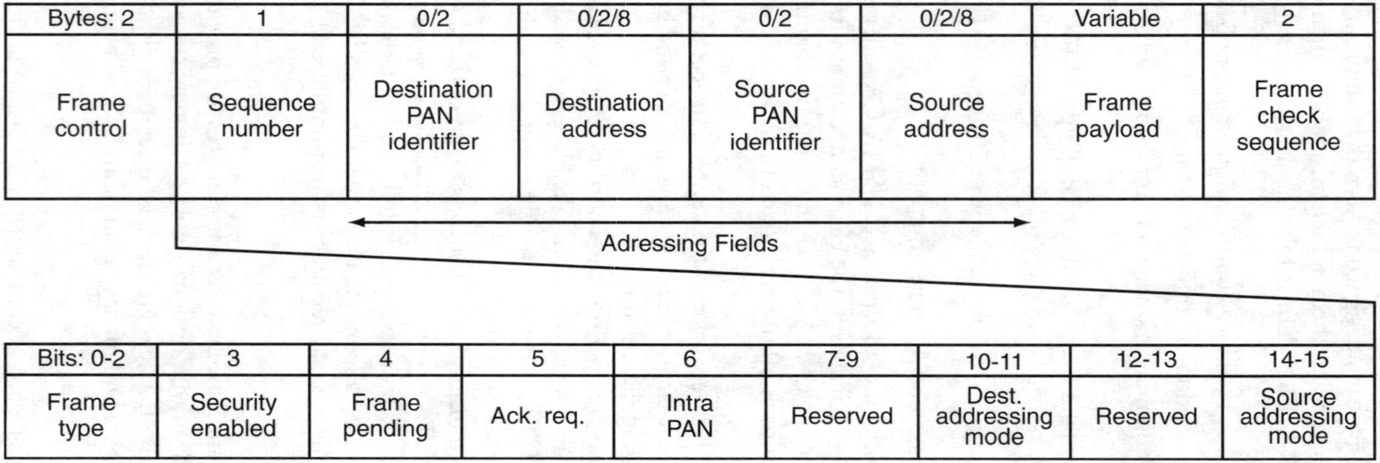

Bytes: 2	1	0/2	0/2/8	0/2	0/2/8	Variable	2
Frame control	Sequence number	Destination PAN identifier	Destination address	Source PAN identifier	Source address	Frame payload	Frame check sequence

Adressing Fields

Bits: 0-2	3	4	5	6	7-9	10-11	12-13	14-15
Frame type	Security enabled	Frame pending	Ack. req.	Intra PAN	Reserved	Dest. addressing mode	Reserved	Source addressing mode

Figure 4.17 802.15.4 frame format.

- *Data encryption:* In 802.11.4, data encryption is based on symmetric keys that are shared by a group of devices (a network or two peers), which can provide protection for beacon, command, and data payloads. The key employs AES with 128 bits, which is currently one of the best encryption schemes.

- *Frame integrity:* Frame integrity uses a message integrity code to protect data from being modified by third parties without the cryptographic key (providing some protection against a man-in-the-middle attack).

- *Sequential freshness:* Sequential freshness is provided by *freshness values*, which are incremented with time and ensure that old frames are not repeated by malicious third parties (providing some protection from the stream-reuse attack). The principle of sequential freshness is simply that an incoming frame must have increasing freshness values; if not, then the incoming frames are discarded.

The three security modes that are provided are as follows:

- Unsecured mode;
- ACL mode, where a higher layer may chose to reject frames based on the ACL list maintained by the MAC layer;
- Secured mode, which provides any of the security features defined above.

4.4.3 The Physical Layer

The physical layer is based on direct sequence spread spectrum communication with BPSK and O-QPSK modulation[1] offering 20- to 250-Kbps rates as defined in Table 4.4.

The 868- and 915-MHz bands, offering 20 and 40 Kbps, use BPSK modulation with differential encoding (for reception simplicity) and use a 15-chip m-sequence as a spreading code. The BPSK chips are filtered with a raised-cosine filter with an excess bandwidth of 100% (hence, 600-kHz bandwidth at 868 MHz and 1,200 KHz at 915 MHz, where the channel separation between the 10 channels is 2 MHz).

The 2.450-GHz band, offering 250 Kbps, uses 16-ary orthogonal data modulation (i.e., 4 bits/symbol, 62.5 Ksps). The symbols are mapped to an

---

1. Offset-QPSK is a QPSK modulation in which the in-phase and in-quadrature components do not change at the same time, but rather with an offset of half a symbol time, yielding less amplitude variation and hence a lower peak to average power ratio. GSM uses a form of O-QPSK.

Table 4.4
Frequency Bands and Data Rates of 802.15.4

PHY (MHz)	Frequency Band (MHz)	Spreading Parameters		Data Parameters		
		Chip Rate (Kcps)	Modulation	Bit Rate (Kbps)	Symbol Rate (Ksps)	Symbols
868/915	868–868.6	300	BPSK	20	20	Binary
	902–928	600	BPSK	40	40	Binary
2,450	2,400–2,483.5	2,000	O-QPSK	250	62.5	16-ary Orthogonal

orthogonal set of 32-chip quasi-orthogonal PN codes, which are O-QPSK modulated with half-sine pulse shaping filtered. The chip rate obtained is thus of 2.0 Mcps, and the channel separation between the 16 channels is 5 MHz. (See Figure 4.18.)

The devices should be capable of transmitting at least –3 dBm; the maximum power is not defined by the standard, but is limited by each country's regulatory limits. The receiver sensitivity is set to –85 dBm for the 2.4-Ghz operation and to –92 dBm for the 868/915-Mhz operation, which are less loose values than for Bluetooth, but still offering space for low-power design.

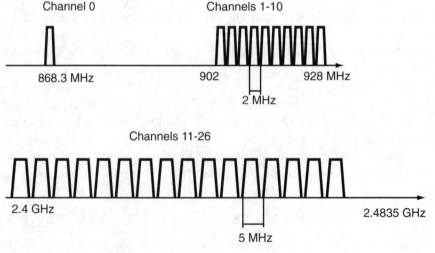

Figure 4.18 802.15.4 channels.

Coexistence with other communications standards is addressed by dynamic channel selection, while packet strength indication is provided mainly for higher layer purposes (e.g., routing).

The physical packet is composed of the following:

- Preamble (32 bits) for symbol synchronization;
- Start of packet delimiter (8 bits) for frame synchronization;
- PHY header (8 bits), which specifies PSDU length (on 7 bits);
- Payload, with up to 127 bytes.

4.4.4 TG4b Development

The 802.15.4b work group, besides working on the natural evolution of and corrections to the original standard, is developing an additional PHY layer, which will mainly provide 250 Kbps at the 868/915-MHz band, either by using 2-MHz bandwidth, with a similar modulation scheme as for the 2.4-GHz band, or by using a 600-kHz bandwidth, with a BPSK/ASK modulation with parallel sequence spread spectrum, providing higher spectral efficiency.

4.4.5 ZigBee

ZigBee is the name of a nonprofit industry consortium consisting of semiconductor manufacturers, OEMs, and solution providers. Its main objective is to define an open global standard for the wireless sensor networking domain, especially for home/building automation and control, medical monitoring, toys, as well as PC-related devices and consumer electronics. ZigBee is based on the 802.15.4 lower layers and adds logical networking, security management, and software, as well as application software.

The promoter companies are Honeywell, Invensys, Mitsubishi, Motorola, Samsung, and Philips. The whole spectrum of targeted applications is well represented by these six companies, and more than 90 companies joined the ZigBee Alliance.

The origin of the word ZigBee comes from the way honey bees communicate the locations of new-found food sources [16]:

"The domestic honeybee, a colonial insect, lives in a hive that contains a queen, a few male drones, and thousands of worker bees. The survival, success, and future of the colony is dependent upon continuous communication of vital information between every member of the colony. The technique that honey bees use to communicate new-found food sources to other members of the colony is referred to as the ZigBee Principle. Using this silent, but powerful communication system, whereby the bee dances in a zigzag pattern, she is able to share information such as the location, distance, and direction of a newly discovered food source to her fellow colony members. Instinctively implementing the ZigBee Principle, bees around the world industriously sustain productive hives and foster future generations of colony members."

4.5   ZigBee and 802.15.4 Products

802.15.4 and ZigBee will most probably provide the next big wave of wireless products in the coming years. Since big home/industrial automation companies like Honeywell and Danfoss are in the alliance, chances are big that ZigBee will be the real move towards the wireless automated home. Analyst Kristen West predicts that there could be more than 50 ZigBee devices in a home, leading to mass productions of up to 150 million devices in 2008 (according to Instat/MDR). This number is impressive but still below the numbers that are predicted for Bluetooth (a little under 1 billion devices).

What follows are the major chips/software solutions for ZigBee/802.15.4.

802.15.4 Chipsets/Systems (Some with Software Protocols)

- Airbee Wireless and Link Plus Corporation codevelop low-cost systems on a chip around the ZigBee and IEEE 802.15.4 wireless standard. The resulting products will apply Link Plus AWICS radios and microprocessor-based hardware and Airbee UltraLite embedded software protocols.

- Atmel offers a two-chip solution, consisting of a transceiver (currently for the 868/915-MHz band) and an 8-bit microcontroller-based controller, to form a complete solution with Figure 8's software (see below).

- Chipcon, based in Norway, provides 2.4-GHz 802.15.4 ZigBee-ready single chip solution (CC2420) and a 868/915-MHz chip (CC1020), as well as support for the ZigBee protocol stack and development tools.

- Ember (www.ember.com, backed by Microsoft) provides a 2.4-GHz 802.15.4 radio chip as well as a 2.4-GHz ZigBee platform based on the CC2420 chip and is IEEE 802.15.4 compliant and ZigBee ready.

- Freescale/Motorola provides a ZigBee-ready platform, currently targeted at development rather that providing a one-chip solution.

- Jennic provides a 2.4-GHz IEEE 802.15.4 transceiver device for IEEE 802.15.4 and ZigBee standards.

802.15.4 Network Solutions (Proprietary)

- EmberNet (provided by Ember Corp.) is a protocol stack developed on top of IEEE 802.15.4 that offers self-healing network functionalities and provides support up to the transport layer as well as security management and APIs for easy implementation.

- Millennial Net, Inc. (www.millenial.net) provides IEEE 802.15.4 modules with a proprietary protocol for mesh/star sensor networks.

ZigBee Network Solutions

- Figure 8 Wireless, Inc. provides a complete implementation of the Zig-Bee Alliance protocol stack, as well as a Windows-based tool to permit viewing of trace and debug information as well as scripting.

- CompXs (www.compxs.com) provides complete 802.15.4-based solutions (integrating Chipcon IP) and chips, implementing ZigBee networking. CompXs also provides 802.15.4 IP cores.

- Helicomm (www.helicomm.com) provides complete ZigBee- and IEEE 802.15.4–based wireless networking solutions, from IPs to hardware and software, selling its services for OEM and systems integrators.

- Crossbow Technology (www.xbow.com) offers 802.15.4 Chipcon-based solutions supporting low-power mesh networking based on U.C. Berkeley's Tiny-OS, which was developed after the SmartDust project. Their latest small mobile terminals are ZigBee-ready.

4.6 Conclusions

For the last 5 years, industry has been putting a lot of effort into WPAN standardization through the IEEE 802.15 group. Recognizing that WPANs will address a large variety of applications (and hence data rates), the standardization efforts have resulted in standards that cover low rate applications (from 1 Kbps, with 802.15.4a) to those with rates up to 1 Gbps (with 802.15.3a), as illustrated in Table 4.5.

These standards have also addressed some shortcomings of the IEEE 802.11 MAC layer, providing the tools for QoS and better security. On the more technological side, it has benefited from the nascent UWB technologies (even if fierce battles have slowed down the adoption process) and has introduced networking technologies for wireless mesh topologies.

As products become available for these standards (and many are on their way, besides the well-established Bluetooth devices), the pervasiveness of communications will become a reality, and the establishment of true personal networks will be at anybody's hand.

Table 4.5
802.15 WPAN Standards

Service	802.15.1	802.15.3	802.15.3a	802.15.4	802.15.4	802.15.4a
Frequency Band	2.4 GHz	2.4 GHz	UWB	896/902 MHz	2.4 GHz	UWB
Data Rate	1 Mbps	11, 22, 33, 44, 55 Mbps	110–480 Mbps	20/40 Kbps	250 Kbps	
Range	10m Class 3 100m Class 1	10m	4.5m > 200 Mbps 10m > 110 Mbps	10m 100m	10m 100m	
Clock Accuracy	± 20 ppm	± 25 ppm	± 25 ppm	± 40 ppm	± 40 ppm	
Current Drain (mA)	<30	<80	30–80 <100	6-mo battery life	6-mo battery life	
Complexity	1	1.5x	2x	0.2x	0.2x	
Connect Time	5 sec	<<1 sec		1 sec to 1 hr	1 sec to 1 hr	
QoS	SCO voice Async data	GTS	GTS	GTS	GTS	GTS
Number of Channels	None (FH)	5	TBD	1/10	16	
Number of Nodes	8 per piconet 64 per scatternet	8-bit piconet address	8-bit piconet address	8- or 64-bit piconet address	8- or 64-bit piconet address	8- or 64-bit piconet address

References

[1]   IEEE 802.15 group, documents available at http://www.ieee802.org/15.

[2]   Karaoguz, J., "High-Rate Wireless Personal Area Networks," *IEEE Communications Magazine*, December 2001, pp. 96–102.

[3]   802.15.3 Standard, "Wireless Medium Access Control (MAC) and Physical Layer (PHY) Specifications for High Rate Wireless Personal Area Networks (WPANS)," IEEE Computer Society, September 29, 2003.

[4]   Gilb, James, P. K., "Wireless Multimedia: A Guide to the IEEE 802.15.3 Standard," IEEE Press, April 2004.

[5]   Callway, E., et al., "Home Networking with IEEE 802.15.4: A Developing Standard for Low-Rate Wireless Personal Area Networks," *IEEE Communications Magazine*, August 2002, pp. 70–77.

[6]   Zheng, J., and M. J. Lee, "Will IEEE 802.15.4 Make Ubiquitous Networking a Reality? A Discussion on a Potential Low Power, Low Bit Rate Standard," *IEEE Communications Magazine*, June 2004, pp. 140–146.

[7]   802.15.4 Standard, "Wireless Medium Access Control (MAC) and Physical Layer (PHY) Specifications for Low Rate Wireless Personal Area Networks (WPANS)," IEEE Computer Society, September 29, 2003.

[8]   Gutierrez, J. A., E. H. Callaway, and R. Barrett, *IEEE 802.15.4 Low-Rate Wireless Personal Area Networks: Enabling Wireless Sensor Networks*, IEEE Press, April 2003.

[9]   Ungerboeck, G., "Trellis-Coded Modulation with Redundant Signal Sets," *IEEE Communications Magazine*, Vol. 25, February 1987, pp. 5–11.

[10]   Milewski, A., "Periodic Sequences with Optimal Properties for Channel Estimation and Fast Start-Up Equalization," *IBM J. Res. Develop.*, Vol. 27, No. 5, September 1983, pp. 426–431.

[11]   Roy, S., et al., "Ultrawideband Radio Design: The Promise of High-Speed, Short-Range Wireless Connectivity," *IEEE Proceedings*, Vol. 92, No. 2, February 2004, pp. 295–311.

[12]   Win, M. Z., and R. A. Scholtz, "Impulse Radio: How it Works," *IEEE Communications Letters*, Vol. 2, No. 2, February 1998 pp. 36–38.

[13]   Engels, M., (Ed.), *Wireless OFDM Systems: How to Make Them Work?* Boston, MA: Kluwer, August 2002.

[14]   van Nee, R. D. J., and R. Prasad, *OFDM for Wireless Multimedia Communications*, Norwood, MA: Artech House, 2000.

[15]   Gupta, P., and P. R. Kumar, "The Capacity of Wireless Networks," *IEEE Trans. on Information Theory*, Vol. 46, March 2000, pp. 388–404.

[16]   ZigBee FAQ, available at http://www.ZigBee.org.

[17]   Porcino, D., and W. Hirt, "Ultra-Wideband Radio Technology, Potential and Challenges Ahead," *IEEE Communications Magazine*, July 2003, pp. 66–74.

Part 2
Trends and Research Topics

5

WPAN Scenarios and Underlying Concepts

According to the IEEE 802.15 standard, "Wireless personal area networks (WPANs) are used to convey information over relatively short distances among a relatively few participants. Unlike wireless local area networks (WLANs), connections effected via WPANs involve little or no infrastructure. This allows small, power efficient, inexpensive solutions to be implemented for a wide range of devices." Bluetooth defines itself a little differently: "Bluetooth wireless technology is a short-range communications system intended to replace the cable(s) connecting portable and/or fixed electronic devices. The key features of Bluetooth wireless technology are robustness, low power, and low cost."

Because Bluetooth defines WPANs as a "mere cable replacement," WPAN applications may function under a very wide range of different operating scenarios, and hence lead to very complex systems, going further in user friendliness than does 802.15.

This chapter presents three classical WPAN scenarios, addressing both the business and the private spheres. These scenarios were conceived and partly written during an internal workshop of the Hermes Partnership, and they led to the definition of WPAN and PN requirements and architectures in PACWOMAN and MAGNET (which will be given in Chapters 6 and 9, respectively). These scenarios will be extended in Chapter 10 in the light of the personal networking paradigm.

The second part of this chapter focuses on two concepts that motivate the use of WPANs and PNs. The first one is the concept of the *disappearing computer*, which has led to new communication needs that can be fulfilled by WPANs/PNs and also help in their design. The second one is the concept of *awareness services*, whose objective is to help people that are far away to keep in touch—actually, *keep aware*, which is much stronger than just keeping in touch—with each other. Both these concepts have triggered many research projects, and these are presented in Chapter 6.

The chapter's third part introduces the specific security issues that will be detailed in Chapter 8.

5.1 Classical WPAN Scenarios

Three key operating scenarios are identified as illustrative of the capabilities of WPANs and as promising from a business perspective: business services, personal services, and health services. From these scenarios, three main categories of applications can be identified, spanning several orders of magnitude in terms of the required information transfer rate: (1) very low speed sensor and control sensor data; (2) low to medium speed medical telemonitoring and business services; and (3) medium to high speed multimedia and interactive video services or time-critical computer data transfers.

In the rest of this section, we provide three simple and typical scenarios. Unlike in [1], which divided the scenarios from the point of view of social needs and implications, we instead take the point of view of a specific person (i.e., a business woman, a person at home, and a heart patient), and follow this person determine what help he or she would want from her environment. We complete these descriptions with a table showing the devices that would be needed here. We do not develop the architecture, as that is done in the next chapter, Section 6.4. To learn how scenarios can be built and lead to relevant system requirements and architecture definitions, we refer the reader to [2], which is a result of the MAGNET project [3]. Other typical scenarios can be found in [1, 4, 5].

5.1.1 The Traveling Businesswoman

Scenario

Mrs. Smith, a British government delegate, is approaching the Waterloo train station. On her way to the station, she speaks into her wearable listening device stating that she wants to visit Mr. Martin. Her WPAN inquires as to Mr. Martin's current location and guides Mrs. Smith to the right train, heading for Paris. Arrow indications are given through her glasses, directing her to and inside the station. Once she is confident she is on the right train, she again speaks into the listening device and requests a number of very confidential video documents, which will help to prepare her for the meeting.

Now satisfactorily prepared, she relaxes in her train seat and is offered a host of different entertainment services. She has no budget limits, so she automatically receives high-quality versions without commercials, and automatically pays with her credit card.

After a short time, a fellow traveler, Mr. Neighbor, talks to her, and they exchange their video business cards. Mr. Neighbor is looking at the latest video clips from a popular music group, and since these were not available on the train's server, Mrs. Smith tunes her glasses and earphones onto Mr. N's mobile device and they share the clips.

At some point, Mrs. Smith's phone rings. Mr. Martin is calling in advance of the meeting and asks her to look at some documents (she can do this through her glasses, but because they are not confidential she chooses to view them on the display provided by the train).

After the meeting in Paris with Mr. Martin, Mrs. Smith takes advantage of Mr. Martin's office. She writes her reports, prints them on the local printer, and leaves the office that evening. She requests her PAN to find a suitable hotel, waits for the taxi her PAN has called for, and enters her room. Check-in and billing is handled by her system. The room area network has been informed of Mrs. Smith's preferences, and it makes a final check of her identity using the latest bio-security devices.

In the morning, before going back to London, Mrs. Smith decides to do some shopping. Beforehand, she requests a remote look at some the most remarkable pieces from various jewelers. Upon entering the black jewel shop, a commercial pops up on her glasses; it is an advertisement for the famous black diamond that shows a small movie on how the bushmen came across this large stone, and the subsequent processes that led to this masterpiece of jewel work. Of course, the commercial agency knows she can afford this purchase (unless she had turned on a privacy profile to avoid this sort of spam).

Comments

This scenario was created after initial work for PACWOMAN [6, 7], and it is quite typical of what a person of business might experience in the (near) future. It is also quite similar to two scenarios envisioned by IST Advisory Group's work on scenarios for ambient intelligence in 2010 [1]. In the scenario described above, the devices used range from very simple sensors (with very low data rate) to high-end devices (with very high data rates and high movement speeds). Mobility is, of course, of prime importance, as is the need to communicate at large distances over some mobile Internet or through cellular mobile networks.

Both 802.15 and Bluetooth WPAN definitions are relevant here for most of the individual situations (in the train, in the office, in the hotel room). When it comes to supporting mobility, however, access to the outer world could be provided by a gateway (see, e.g., Section 6.4), while services could be provided by a third party service provider. Obviously, to provide a seamless service, the classical WPAN must be complemented with mobility support (which was provided in PACWOMAN) and service support.

5.1.2 The Intelligent House

Scenario

At five o'clock in the evening, in the Paris suburbs, Mr. Wood's family members are heading towards home. Mrs. Wood's last meeting has concluded and she asks for a taxi, indicating her home address. The livery company, automatically recovering information on who she is and her transportation habits, quickly sends her an electric bike-like vehicle, the only way to assure her on-time arrival at home. In the meantime, her husband has proposed the menu for the dinner meal; they discuss it briefly as he heads from work to the market. The home computer tells him what to buy according to his menu and what they still have at the house. Although the house is mostly automated and adapts to the family schedules, Mr. Wood takes a look at what is happening around and inside his house through his car's display and checks the temperature.

The children have finished school and go home with some friends. They retire to their bedroom, where the wall display instantly goes on. They first choose to look at a short excerpt of the movie Skywalker, while they call other friends and prepare themselves to play video games [8]. The Skywalker movie is physically located somewhere in the home system; it is transmitted through the high-speed wired backbone, and for the journey from the room's wall to the actual display, it takes a wireless path. Once their remote friends are ready to play, they form a private and secured network, using highly interactive video and sound, even though they cannot afford the higher end 3D gaming options.

Comments

Quite surprisingly, the first combined appearance of the words "wireless" and "intelligent house" appeared in 1995 [9]. Ten years later, the concept of an intelligent house equipped with a wireless network to support this intelligence appears as a most natural idea; the concept of ambient intelligence has rapidly drifted from the business-oriented sphere to the personal sphere. Indeed, it is hard to track papers by focusing only on the (wireless) intelligent house, even if there are thousands of papers citing the term. (A special issue of *IEEE Wireless Communications* introduces the topic [10].) From a general perspective, the real challenge is to make all these appliances *invisible* to the user; and we refer the interested reader to [11], which is a collection of papers on the *invisible future* that addresses issues beyond technicalities, and to [12], which addresses both technological and social aspects of smart environments.

Focusing on the role of WPANs, this scenario is probably the most demanding, both in terms of transmission speed and QoS (in particular, the time constants have to be very low). For multimedia applications, which are prevalent in the home environment, broadband (i.e., an aggregate throughput up to the gigabit per second range) is absolutely necessary; and for gaming applications, high throughputs with very short delays are wanted. The low data rate part of WPANs is also very important for automation purposes. Last but not least, the user would never accept this wireless assisted intelligent home without protecting privacy and ensuring security (or at least they should never accept it, as current experience shows that users are not sufficiently security-aware).

5.1.3   Mobile Health

Scenario

Mr. Graybeard is an elderly person who is under continuous medical telemonitoring because of his health condition. He wears medical sensors and other devices that interact within his room, but also throughout and even outside of his house. There are medical telemonitoring devices for in-home monitoring, as well as for out-of-home monitoring during outside activities. Mr. Graybeard's sensors are for monitoring insulin levels, blood pressure (taken when he is still), temperature, heart rate, EKG, and alarms (e.g., fall detection). In case of critical situations, alarms may be activated and transmitted to the appropriate healthcare professionals. In other cases, medication may be automatically released, or even injected into his body.

Mr. Graybeard lives in a smart home equipped for in-home monitoring and control. He enters his home and checks the sensors, which include preferences for radio, television, as well as alarms to indicate the loss of an item (e.g., keys) or various conditions (e.g., empty refrigerator, low fuel level in the home heating system, an open door). Some of these sensors may be fixed (located on the ceiling, for example).

Mr. Graybeard's wife, who also lives in the house, can be informed of potentially dangerous health situations in order to help with timely intervention, if necessary. She can also check certain health parameters through her husband's wearable devices. For security reasons, they may need to authorize each other in order to exchange confidential data (e.g., health results, payment information). Both of them may wish to send the measured parameters to a doctor, in order for medication to be prescribed.

Mr. Graybeard recently underwent very complicated heart surgery, and his doctor, who is in another town, wants to check his condition and wants to be informed about any irregular situations. This is done remotely. His other health care providers may also be notified in the event of an alarm, indicating that a critical threshold has been exceeded and a potential medical emergency exists. Furthermore, sensor data may be transmitted in the form of regular updates. A wide range of personal medical parameters may be transmitted and stored in a remote medical database, enabling health care providers, with the appropriate access, to periodically check on the patient's status. Additionally, two-way voice service is envisioned to help reassure the patient in the event of a problem. Data service is also two-way, to allow health care providers to request updates of specific parameters.

If the situation becomes too serious, and Mr. Graybeard is unable to express himself, the doctor can potentially view the patient through a small camera. He can then contact another doctor, in the same or another hospital, for advice and exchange of data about the patient's condition or his medical history.

Several weeks after the surgery, Mr. Graybeard feels very bad during his walk in a visiting town. He urgently needs to know the address of the nearest hospital and information on how to get there. He announces his visit to the available doctor, and briefs him on his medical history, while the doctor can see the medical detail on his own display. The hospital checks his medical insurance file.

Comments

The services considered in the telemonitoring and smart healthy home scenario include both data and voice (medium rate PAN scenario), as required to support in-home and mobile medical telemonitoring. Although not specifically addressed in this scenario, advanced telemonitoring methods may also incorporate visual results (i.e., higher speed video service). With medical care costs on the rise and people living longer, in-home patient care is becoming increasingly important. Many elderly people require only occasional monitoring or medical support in order to continue to live comfortably at home. There are potentially tens of millions of medical telemonitoring service users, which makes the ability to support reliable, secure, and scalable medical telemonitoring an important service for future PANs.

Interestingly, the literature on these topics falls into two main categories. The first is devoted to the actual monitoring sensors (e.g., [13, 14]) and decision systems (e.g., [15, 16]). The second is devoted to communications systems that make the above scenario possible (e.g., [17–23], an outcome of the EU URSAFE project). An interesting overview of activities in the field of mobile health is proposed by Istepanian, Jovanov, and Zhang in [24].

From Figure 5.1, it is clear that providing a mobile health service has requirements on low power (for the sensors and the personal server) as well as access to a mobile network infrastructure. Moreover, the scenario is quite demanding in terms of QoS (especially delay).

This scenario shows that the low-power and robustness characteristics of WPAN are indeed crucial and that the ease of use introduced by Bluetooth is

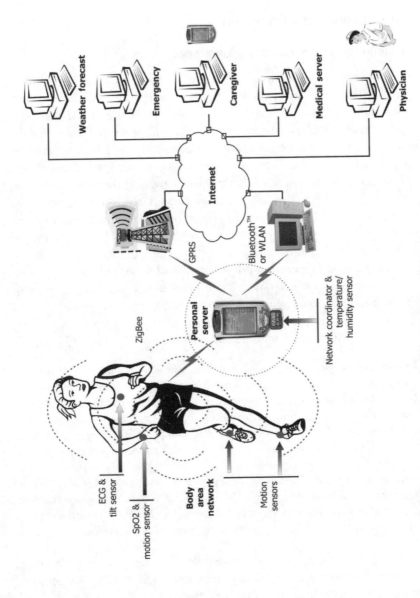

Figure 5.1 A typical mobile health scenario. (*From:* [21]. Reprinted with permission.)

necessary. But WPAN, defined as in the introduction, does not suffice to provide adequate mobile health service, as this service will rely on an integrated network, which is one of the motivations of the PN concept.

5.2 The Disappearing Computer

5.2.1 Invisibility and Technology Unawareness

The WPANs described in Section 5.1 concentrate on the wireless communication aspects, while Bluetooth pays attention to the devices' ease of use (by replacing the cable clutter with wireless technology). Invisibility (of the wires) is the main motto of WPAN. Unawareness, a similar concept, is probably what the user needs. Indeed, a wired phone can be considered an "invisible" technology, due to its ease of use. Technology unawareness refers to this kind of invisibility, as the phone has become so natural that one is not aware of the technological aspects that are behind it. Technology unawareness is also the main motto of Bluetooth in the WPAN community, as Bluetooth is concerned with cable replacements and is eager not to introduce additional complexity for the user (if you remove the wire, but add lengthy and tedious configuration processes, technology unawareness disappears, and the original goal is not met).

Technology unawareness is wanted in other domains, like computing, as the computer is a piece of technology that is complex for the user. Applied to computing, technology unawareness is called *ubiquitous computing*. In the late 1980s, Mark Weiser defined the main ideas behind ubiquitous computing [25–27]:

> Inspired by the social scientists, philosophers, and anthropologists at PARC, we have been trying to take a radical look at what computing and networking ought to be like. We believe that people live through their practices and tacit knowledge so that the most powerful things are those that are effectively invisible in use. This is a challenge that affects all of computer science. Our preliminary approach: Activate the world. Provide hundreds of wireless computing devices per person per office, of all scales (from 1″ displays to wall sized). This has required new work in operating systems, user interfaces, networks, wireless, displays, and many other areas. We call our work "ubiquitous computing." This is different from PDAs, dynabooks, or information at your fingertips. It is invisible, everywhere computing that does not live on a personal device of any sort, but is in the woodwork everywhere.

In Europe, a large research initiative called The Disappearing Computer [28] explored how the use of a network of interacting and invisible devices could enhance a person's daily life. This initiative supported 17 different projects, with a (small) number of them having follow-ups in the Sixth Framework Program of the European community (EC). A short overview of the outcome of these projects can be found in [29] and companion papers in the same journal.

The main challenge in this concept is that the user does not want to be aware of the technology, hence the awareness has to shift from the human to the machine and the environment has to adapt automatically to the person (which is somewhat the contrary to virtual reality: the system has to disappear into the reality, not create another world). Hence, the technologies have to be context aware and eventually have to support social awareness.

5.2.2 From Context Awareness to Social Awareness

As described in the previous section, Bluetooth's goal of cutting the wires can be translated to the user's request to be unaware of the technology. The immediate counterpart of this is that technology has to be aware of the user's need and hence of its context: this is called context awareness. On the other hand, the goal of any means of communication is for a person to be in touch with another person. This goal has been splendidly fulfilled both by the classical telephone and the mobile telephone. This personal communication is the primitive form of social awareness.

Context awareness is one of the keys of pervasive computing. Coutaz et al. [30] introduce the concept as follows: "Context is not simply the state of a predefined environment with a fixed set of interaction resources. It's part of a process of interacting with an ever-changing environment composed of reconfigurable, migratory, distributed and multiscale resources."

From a purely technical point of view, context awareness of services consists of location, time, user role, knowledge for environment restrictions, available capabilities or services, and device awareness. The implication of this on WPANs is the need for heterogeneous networks (i.e., wired/wireless, cellular/ad hoc/WLAN-like, and so forth) and the need of a large range of different devices. Moreover, context awareness has given birth to whole research fields like localization, service discovery, and context identification, which are definitely part of the WPAN paradigms (see, for example, the Service Discovery Protocol in Bluetooth).

Both in the private and business spheres, social awareness is probably one of the most challenging and desirable features of personal assistants, and hence, the design of WPANs must support the social awareness systems.

According to [31], "Awareness systems are computer mediated communication (CMC) systems that aim to support connected parties to maintain a peripheral awareness of the activities and status of their communication partners. Awareness systems can serve both work-related and social interactions." Social awareness is also a research field in itself (to which WPAN design is attentive), one that has many technical implications on both devices (simple, unobtrusive) and on human–environment interaction (no longer merely human–machine interaction), which in the end is a human–human computer-mediated interaction. Most of the work on social awareness of system design (see, for example, [28,

32], as well as the journal *ACM Transaction on Computer-Human Interactions and ACM Interactions*) is devoted to the devices and to the interfaces, while others look at the convergence of current wireless technologies (and the search for new markets).

For WPAN, social awareness is the ultimate condition for success (as it was for telephony) and will be the "killer application" for the future wireless industry.

5.3 Security Aspects

The main concerns of WPAN users are privacy and trust issues (see [33, 34]). Security threats on wireless networks are quite well known (see Chapter 8), but the shift to WPANs and pervasiveness puts the user to the test.

Trust. Suppose you want to buy a train ticket while you are in a grocery store and you do this by the means of your WPAN. You need a high level of trust in the system to be sure about the identity of the issuer. This is a classic trust issue that can be solved with current security techniques, even if the end-to-end security still has to be proven in the WPAN context.

Privacy. Suppose now that you have an argument with you favorite grocery clerk and need to go to court (which is a true example given in [33]). You have always used your bank card to pay, and the shop has recorded everything you have purchased. Based on these records, the shop can denounce your alcoholic habits and use that in their legal defense. This is one of the numerous examples of privacy issues. These privacy issues are linked to the fact that smart environments are constantly sensing and monitoring your activities, they have a great deal of information about you, and hence, they can violate your privacy very deeply (George Orwell's *1984* is probably the best description of these threats). What is troubling, however, is that, according to [33], most researchers in these fields are not concerned so much about these issues and believe that the application of classical security receipts will solve the problem; although Lahlou et al. propose some guidelines in [35].

Another distinct feature of WPANs is the presence of sensor networks and/or radio frequency identification (RFIDs). Both their low-power aspects and the potentially large number of sensors pose different problems [i.e., the distribution of certificates, the use of low-power algorithms (hence, cryptographically weak), openness to attacks on routing, and the ease of node falsification]. The RFID's main problem, from the perspective of the user, is a privacy issue [34], although falsification is the main concern for the manufacturer. In sensor networks [36], besides the possible use of a base station as a point of trust (meaning that it can also be attacked by impersonation), the main problem is the introduction of malicious nodes that can implement several attacks (denial of service by jamming and packet injection, claiming of multiple identities and routing attacks by dropping packets, and spreading bogus routing information and flooding).

Solutions to these problems include the implementation of secure routing (by multipath routing on secured links), secure localization, code attestation (as compromised nodes usually run code that is different from an authorized node), and the design of efficient cryptographic primitives for low-power operation.

References

[1]   Ducatel, K., et al., (Eds.), "Scenarios for Ambient Intelligence in 2010," IST Advisory Group (ISTAG), European Commission, Brussels, 2001.

[2]   Schultz, N., D. Saugstrup, and L. Sørensen, "A User-Centred Scenario Framework Using Creative Workshops," *Proceedings of the Fourth Danish Human-Computer Interaction Research Symposium*, Aalborg University, Denmark, November 16, 2004, pp. 5–8.

[3]   MAGNET project, available at http://www.ist-magnet.org.

[4]   mITF, "Flying Carpet, Towards the 4th Generation Mobile Communications Systems," v2.00, 2004, available at http://www.mitf.org/public_e/archives/Flying_Carpet_Ver200.pdf.

[5]   Batteram, H., (Ed.), et al., "Scope and Scenarios," Deliverable D1.1 of the Freeband/ Awareness Project, October 2004, available at http://www.freeband.nl.

[6]   European IST project, "Power Aware Communications for Wireless Optimised Personal Area Networks," Contract No IST-2001-34157, available at http://www.imec.be/pacwoman.

[7]   Louagie, F., L. Munoz, and S. Kyriazakos, "Paving the Way for Fourth Generation: A New Family of Wireless Personal Area Networks," IST Mobile Summit 2003.

[8]   Stone, A., "Wireless Makes Multiplayer Gaming a Winner," *Pervasive Computing*, IEEE, Vol. 2, No. 4, October/December 2003, pp. 5–9.

[9]   Takada, H., and K. Sakamura, "Compact, Low-Cost, but Real-Time Distributed Computing for Computer Augmented Environments," *Fifth IEEE Workshop on Future Trends of Distributed Computing Systems*, Chenju, Korea, August 28–30, 1995, p. 56–65.

[10]   Das, S. K., and D. J. Cook, "Guest Editorial—Smart Homes," *Wireless Communications*, IEEE, Vol. 9, No. 6, December 2002, pp. 62–62.

[11]   Denning, P. J., (Ed.), *On the Seamless Integration of Technology into Everyday Life—The Invisible Future*, New York: McGraw-Hill, 2002.

[12]   Cook, D. J., and S. K. Das, (Eds.), *Smart Environments*, Wiley-Interscience Series in Discrete Mathematics and Optimization, New York: John Wiley & Sons, Inc., 2005.

[13]   Korhonen, I., J. Parkka, and M. V. Gils, "Health Monitoring in the Home of the Future," *IEEE Eng. Med. Biol. Mag.*, Vol. 22, No. 3, May/June 2003, pp. 66–73.

[14]   Alcock, S. J., "Technology for Continuous Invasive Monitoring of Glucose," in *Proc. IEEE Conf. Engineering in Medicine and Biology*, Amsterdam, October/November 1996, pp. 2156–2158.

[15]   Asada, H. H., et al., "Mobile Monitoring with Wearable Photopletismographic Biosensors," *IEEE Eng. Med. Biol. Mag.*, Vol. 22, No. 3, 2003, pp. 28–40.

[16]   Amigoni, F., et al., "Anthropic Agency: A Multiagent System for Physiological Processes," *Artif. Intell. Med.*, Vol. 27, No. 3, March 2003, pp. 305–334.

[17]   Chu, Y., and A. Ganz, "A Mobile Teletrauma System Using 3G Networks," *IEEE Transactions on Information Technology in Biomedicine*, Vol. 8, No. 4, December 2004, pp. 456–462.

[18]   Salvador, C. H., et al., "Airmed-Cardio: A GSM and Internet Services-Based System for Out-of-Hospital Follow-Up of Cardiac Patients," *IEEE Transactions on Information Technology in Biomedicine*, Vol. 9, No. 1, March 2005, pp. 73–85.

[19]   Rasid, M. F. A., and B. Woodward, "Bluetooth Telemedicine Processor for Multichannel Biomedical Signal Transmission via Mobile Cellular Networks," *IEEE Transactions on Information Technology in Biomedicine*, Vol. 9, No. 1, March 2005, pp. 35–43.

[20]   Jovanov, E., et al., "Stress Monitoring Using a Distributed Wireless Intelligent Sensor System," *IEEE Eng. Med. Biol. Mag.*, Vol. 22, No. 3, May/June 2003, pp. 49–55.

[21]   Jovanov, E., et al., "A Wireless Body Area Network of Intelligent Motion Sensors for Computer Assisted Physical Rehabilitation," *Journal of NeuroEngineering and Rehabilitation*, Vol. 2, No. 6, March 1, 2005, available at http://www.jneuroengrehab.com/content/2/1/6.

[22]   Stanford, V., "Using Pervasive Computing to Deliver Elder Care," *Pervasive Computing*, IEEE, Vol. 1, No. 1, January/March 2002, pp. 10–13.

[23]   Henrion, S., C. Mailhes, and F. Castanié, "Transmitting Critical Biomedical Signals over Unreliable Connectionless Channels with Good QoS Using Advanced Signal Processing," *Eighth WSEAS International Conference on COMMUNICATIONS*, Vouliagmeni, Athens, Greece, July 12–15, 2004, published in WSEAS Transactions on Communications, Vol. 4, No. 1, July 2004.

[24]   Istepanian, R. S. H., E. Jovanov, and Y. T. Zhang, "Introduction to the Special Section on M-Health: Beyond Seamless Mobility and Global Wireless Health-Care Connectivity," *IEEE Transactions on Information Technology in Biomedicine*, Vol. 8, No. 4, December 2004, pp. 405–414.

[25]   "Ubiquitous Computing," available at http://www.ubiq.com.

[26]   Weiser, M., "The Computer for the 21st Century," *Scientific American*, Vol. 265, No. 3 September 1991, pp. 94–104.

[27]   Weiser, M., "Some Computer Science Issues in Ubiquitous Computing," *Commun. ACM*, Vol. 36, No. 7, August 1993, pp. 74–84.

[28]   "The Disappearing Computer," EU research initiative, available at http://www.disappearing-computer.net.

[29]   Special issue on "The Disappearing Computer," *Communications of the ACM*, Vol. 48, No. 3, March 2005, pp. 33–71.

[30]   Coutaz, J., et al., "Context Is Key," *Communications of the ACM*, Vol. 48, No. 3, March 2005, pp. 49–53.

[31]   ASTRA IST-project, available at http://www.presence-research.org/Astra/index.html.

[32] Riva, G., F. Davide, and W.A. Ijsselsteijn, *Being There: Concepts, Effects and Measurements of User Presence in Synthetic Environments*, Amsterdam: IOS Press, 203.

[33] Lahlou, S., M. Langheinrich, and C. Röcker, "Privacy and Trust Issues with Invisible Computers," *Communications of the ACM*, Vol. 48, No. 3, March 2005, pp. 59–60.

[34] Weiss, S., "Security and Privacy Issues in RFIDs," M.S. Thesis, MIT, Cambridge, Massachusetts, 2003.

[35] Lahlou, S., and F. Jegou, "European Privacy Design Guidelines for the Disappearing Computer," available at http://www.rufae.net/privacy.html.

[36] Shi, E., and A. Perrig, "Designing Secure Sensor Networks," *IEEE Wireless Communications*, Vol. 11, No 6, December 2004, pp. 38–43.

Selected Bibliography

Bardram, J. E., and T. R. Hansen, "Social Awareness and Availability: The AWARE Architecture: Supporting Context-Mediated Social Awareness in Mobile Cooperation," *ACM Conference on Computer Supported Cooperative Work*, November 2004, pp. 192–201.

Berezdivin, R., R. R. Breinig, and R. Topp, "Next-Generation Wireless Communications Concepts and Technologies," *IEEE Commun. Mag.*, Vol. 40, No. 3, March 2002, pp. 108–116.

Bertamini, F., et al., "Olympus: An Ambient Intelligence Architecture on the Verge of Reality," *Proc. Conf. Image Analysis and Processing*, Mantova, Italy, September 2003, pp. 139–144.

Cheverst, K., et al., "Exploiting Context to Support Social Awareness and Social Navigation," *SIGGROUP Bull.*, ACM, Vol. 21, No. 3, 2000, pp. 43–48.

Evans, D., "In-Home Wireless Networking: An Entertainment Perspective," *Electronics & Communication Engineering Journal*, Vol. 13, No. 5, October 2001, pp. 213–219.

Fujieda, H., et al., "A Wireless Home Network and Its Application Systems," *IEEE Transactions on Consumer Electronics*, Vol. 46, No. 2, May 2000, pp. 283–290.

Haas, Z. J., "A Communication Infrastructure for Smart Environments: A Position Article," *IEEE Pers. Commun.*, Vol. 7, No. 5, October 2000, pp. 54–58.

Kulkarni, A., "Design Principles of a Reactive Behavioral System for the Intelligent Room," *Bitstream*, April 2002.

Schilit, B. N., D. M. Hilbert, and J. Trevor, "Context-Aware Communication," *Wireless Communications*, IEEE, Vol. 9, No. 5, Octber 2002, pp. 46–54.

Stanford, V., "Pervasive Computing Puts Food on the Table," *Pervasive Computing*, IEEE, Vol. 2, No. 1, January/March 2003, pp. 9–14.

Streitz, N., and P. Nixon, "The Disappearing Computer," *Communications of the ACM*, March 2005, Vol. 48, No. 3, pp. 33–35

Villar, N., et al., "Interacting with Proactive Public Displays," *Comput. Graph.*, Vol. 27, No. 6, December 2003, pp.849–857.

6

From Pervasive Computing to Personal Networks: A Research Perspective

6.1 Introduction

The concept of WPANs was born with Bluetooth at the end of the 1990s, and it focused on cable replacement and wireless communications for low power and low rate devices. Following Ericsson's lead, a host of companies and research institutes embraced similar ideas and began low power, low rate wireless networks projects. Section 6.2 will detail some of the most relevant of these projects.

The goal of pervasive computing (or ubiquitous computing [1]), as stated in Chapter 5, is a primary motivation for the building of WPANs. Enabled by the advent of embedded technology, the objectives of pervasive computing are (1) to create an environment where the connectivity of devices is embedded in such a way that the connectivity is unobtrusive and always available, and (2) to create an environment saturated with computing devices that is gracefully integrated with human users.

Ambient intelligence adds another aspect, and its goal can be defined: to create an environment to which human users are connected by intelligent user interfaces. This does not only refer to the classical human/machine interface (sensors, actuators, keyboards, displays, speech recognition), but also to the fact that the environment has insight into the way users like to interact with (computing) devices [2].

Wireless networking communication was confronted with the slow start of third generation wireless mobile networks as well as with the explosion of high rate wireless local area networks. These changes blurred the classical mobile communication view, leading to research on new wireless network architectures focusing, like in the pervasive computing and ambient intelligence concepts, on user-centric approaches. Hence, new research projects looked at architectures that would support both ambient intelligence and person-centric networks. In this chapter, we first give an overview of WPAN projects (Section 6.2):

- Initial WPAN projects focusing on the quest for low power;
- Current building blocks used toward the design of ultra low energy network nodes, including energy scavenging, the use of RF-MEMS, and the introduction of ultra-wideband communication technology and cross-layer optimization (whereby communication is optimized by having the different networking layers exchange more information than is currently the case);
- Projects introducing ad hoc networking to provide mobile seamless access.

In Section 6.3 we elaborate on OXYGEN, one of the major projects promoting and developing ambient intelligence, and in Section 6.4 we show the impact of this concept on two European projects developing WPANs (PACWOMAN) and introducing the personal network concept (MAGNET: My Adaptive Global NETwork).

6.2   Overview of WPAN Projects

This section provides an overview of the most relevant projects related mainly to pure WPAN. We provide only a small amount of project descriptions, focusing on the concepts developed. Of course, many other very good research projects exist, but an exhaustive catalog of WPAN projects is beyond the scope of this book.

6.2.1   The Quest for Low Power: Building a Wireless Sensor

Providing the last mile of wireless communication has been a big quest of the last decade that has led to a host of products, notably WLANs. Providing the last inch of wireless communication and enabling real pervasive and embedded communications is the realm of WPANs. As communications must be present everywhere, a user should be able to easily link any device with a wireless extension—this extension should be inexpensive (by 2010, less then 0.10 cents). In the context of wireless sensors, this wireless extension should not rely on an external energy supply, but be able to live its life with the initial energy (battery) it is given.

The first projects that worked on this are those related to sensor networks. We have selected for discussion three major U.S. projects focusing on low-power wireless networks: PicoRadio [3, 4], μ-Adaptive Multi-Domain Power-Aware

Sensors (μAMPS), and SmartDust. All three projects were born in the context of wireless sensor networks, with low-power and large (100–1,000 node) networks in mind. We first give a short overview of the projects themselves, with pointers to relevant papers, and then give an overview of their concepts and achievements (and others not cited).

6.2.1.1 PicoRadio: Ultra Low Energy Wireless Sensor Nodes

PicoRadio, a project hosted by the University of California, Berkeley, develops and implements technologies to enable the building of ultra low energy wireless networks. Its charter is to [3, 4]:

- Develop meso-scale low cost (< 50 cents) transceivers for ubiquitous wireless data acquisition that minimizes power/energy dissipation:
  - Minimize energy (< 5 nJ/(correct) bit) for energy-limited source;
  - Minimize power (< 100μW) for power limited source enabling energy scavenging.
- By using the following strategies:
  - Self-configuring networks;
  - Fluid trade-off between communication and computation;
  - Integrated system on chip (SOC) approach, aggressive low-energy architectures and circuits.

The idea behind this (initial) charter is to enable the advent of large sensor networks. Its application focus has shifted towards the ambient intelligence concept, as well as towards PAN-like applications described in Chapter 5.

6.2.1.2 μAMPS

Much like PicoRadio, MIT's μAMPS project focuses on low power sensor networks, with topics like energy scavenging and low power specific solutions. One of the major differences in their vision is that [28]:

> Our research will focus on innovative energy-optimized solutions at all levels of the system hierarchy including: physical layer (e.g., transceiver design), data link layer (packetization and encapsulation), medium access layer (multi-user communication with emphasis on scalability), network/transport layer (routing and aggregation schemes), session/presentation layer (real-time distributed Operating System (OS)), and application layer (innovative applications). We will investigate techniques to optimize for energy efficiency vertically across the protocol stack.

Besides work on energy scavenging [5] and other more hardware-oriented work, μAMPS concentrates on what is now known as cross-layer optimization; that is, working on all layers [6, 7].

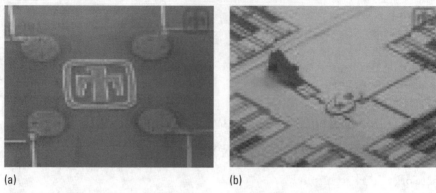

(a)   (b)

Figure 6.1 (a) Four optical encoder wheels. Each wheel has four positions (00, 01, 10, and 11), which can be switched mechanically. One of four signals from fixed-position lasers will be reflected, based on the position of the main encoder wheel. (b) Silicon mirror and drive system: a mirror system design where the mirror is elevated by a three-gear torque-multiplying system. The mirror is shown in the upright position. [Courtesy of Sandia National Laboratories, Summit™ Technologies, http://www.mems.sandia.gov]

6.2.1.3   SmartDust: Communicating with a Cubic Millimeter Computer

The SmartDust project, hosted by the University of California, Berkeley, and finished in 2001, explored "microfabrication technology's limitations to determine whether an autonomous sensing, computing, and communication system could be packed into a cubic-millimeter mote [mobile terminal] to form the basis of integrated, massively distributed sensor networks"[8]. SmartDust concentrated on (Figure 6.1):

- Microfabrication technology on microelectromechanical systems (MEMs), using classical MEMs components as capacitors for electronic systems miniaturization, and also to build microrobots [8];
- Low power computing, using low voltage supply and small size transistors;
- Optical communication and optical sensors.

SmartDust was followed by a collection of projects, now grouped under the Berkeley Wireless Embedded Systems (WEBS) umbrella, and globally focused on developing an open platform for a network of embedded systems, hence broadening the scope of the original project.

6.2.2   (Ultra) Low Energy Network Nodes: How to Get There

The three projects introduced above show us how to enable the design of low energy network nodes. The main concepts are as follows:

- Portable energy sources and energy scavenging [5, 9];
- MEMS for low power and integration;
- Ultra-wideband technology [10];
- Cross-layer optimization [6, 7], which includes:
  - Adaptive technologies (modulation and rate, but also adaptive computation power [6], supply voltage, and computation/communication trade-off);
  - Ad hoc networking as a way to lower energy consumption.[1]

6.2.2.1   Portable Energy Sources and Energy Scavenging: When Nature Provides the Energy [11]

The Holy Grail [3] for low energy sensor networks is for the nodes to be self-contained and self-powered using energy extracted from the environment, which is the definition of energy scavenging.

Let us first review the necessary computation and communication energy. The energy cost of computation with current technologies goes from 0.1 pJ/op[2] (custom devices) up to 100 pJ/op for a microprocessor. Assuming reconfigurable hardware (using a microprocessor is obviously inefficient, but one needs reconfigurable hardware to enable adaptive technology), one can count on about 1 pJ/op, which provides 1 million operations (MOP)/μW. The energy cost of communication, if PicoRadio's charter is attained, is 5 nJ/bit, which provides 200 bits/sec/μW (for a distance of 10m). Hence, if energy scavenging provides 100 μW, a wireless node providing 20 Kbps with limited computing power could be designed, which is a valid target for the industry (this corresponds approximately to IEEE 802.15's lowest rate).

Having these numbers in mind, we can look at natural sources of energy:

- *Solar energy* is commonly available and provides a very good power density (150 μW/cm^2 for a cloudy day, 15,000 μW/cm^2 in direct sun, and about 6 μW/cm^2 for an office desk [11]). The big advantage of solar power is its good power density and availability; the disadvantage is that

---

1. It is important to note that although we are using the term "ad hoc," its meaning for us is limited to the WPAN environment, which is not the same as that used by the IETF Mobile Ad Hoc Networking (MANET) group for much more general networks with high degrees of flexibility and mobility. This deliberate choice is crucial in achieving a reasonable network capacity, which is known to decrease with the square root of the number of active devices in an ad hoc network. This also implies that our solutions to the above concepts must be easier to implement and close to practical infrastructures as, for example, Bluetooth.

2. "op" stands for operation, that is, a basic operation computed on a processor or on an arithmetic and logic unit.

it will not work in the absence of light, so for many embedded devices, other energy sources must be found.

- *Temperature gradients*, as well as daily temperature variations, can be captured by MEMS to provide energy. The basic principle is based on the thermocouple. A thermocouple is a junction built with two different materials (silicium and germanium, for example). When a different temperature is applied to both sides of the junction, a voltage appears at the two poles and can be used to generate energy. To generate a substantial amount of power (around 20 μWatts/cm^3), a collection of thermocouples is brought together with micromachining techniques.

- *Vibrations* can be converted to energy by three means:

    1. *Piezoelectricity:* strain in piezoelectric material yields a voltage across a capacitor [Figure 6.2(a)].

    2. *MEMs capacitor:* in Figure 6.2(b), the capacitor formed by the combs will vary, as the combs move with vibration.

    3. *Induction* of a current in the wire will be yielded by the movement of the coil through the magnetic field in Figure 6.2(c).

    (According to [11], piezoelectric converters appear to be the most promising, with a production of approximately 200 μW/cm^3 against 100 μW/cm^3 for MEMs variable capacitors, using 2,25 m/s^2 vibrations at 120 Hz—which are the vibrations produced by a microwave oven.)

- *Acoustics* are hardly usable, with 3 to 1,000 pW/cm^2.

- *Batteries*, the classic solution, provide up to 1.5 Wh/cm^3, hence providing about 180 μW/cm^3 for a 1-year lifetime and 18 μW/cm^3 for a 10-year lifetime, which is still a good solution for relatively short lifetimes.

- *Fuel cells*, whose availability is now nearly guaranteed, provides 560 μW/cm^3 for a 1-year lifetime, which makes it a good candidate for simple systems, especially because recharging the fuel cell is easy.

This short overview of energy sources shows that it is indeed possible to build self-powered sensor radio nodes, as was demonstrated by [9], which describes a 1.9-GHz radio node powered by a small solar cell and vibration energy scavenging.

6.2.2.2 Radio Frequency Microelectromechanical Systems for Low Power and Integration

Radio frequency microelectromechanical systems (RF-MEMS) are a key element towards low power design and integration of wireless devices. Indeed, in current RF systems, to achieve due precision on oscillators and filters, one is usually forced to use external components, like on-chip analog devices that offer

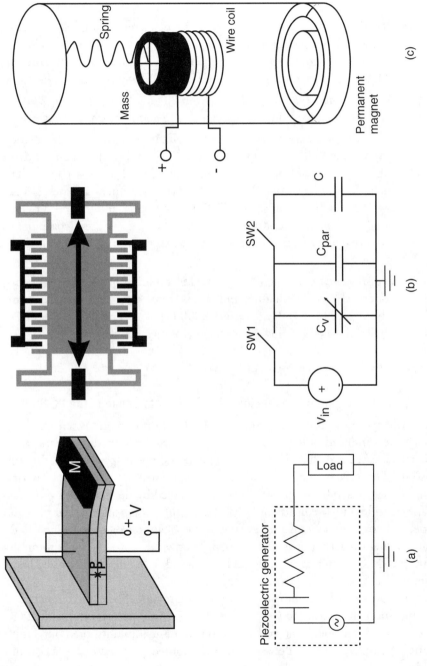

Figure 6.2 Vibrations can be converted to energy: (a) piezoelectricity; (b) MEMs capacitor; and (c) induction.

poor quality factors. In contrast, micromachined components offer high quality factors and enable a drastic miniaturization and integration of the whole transceiver on a single chip. Major components include the following:

- *MEMs inductors* have much lower parasitic elements and self-resonance frequency, hence attaining high quality factors (Qs) that are close to their off-chip counterparts. Qs ranging from 6 to more than 30 are available at frequencies from 2.4 to 18 GHz and inductor values around 1 nH.

- *MEMs tunable capacitors* offer Qs from 20 (at 1 GHz) to 12 (at 2 GHz), with rather large tuning voltage ranges and good linearity characteristics.

- *MEMs switches* are appreciated for their high linearity, high isolation (more than 40 dB at 1 GHz), and low insertion loss (less than 0.1 dB at 1 GHz). The control current is also quite low (less than 10 μA). The main disadvantages of MEM switches for the moment are their switching speed, which is on the order of microseconds, and for specific components their high control voltage (up to 30V, as these components are actuated by electrostatic charges, which have to be large enough for actuation).

- *MEMs resonators*, which can be used as building blocks for oscillators, voltage control oscillators, and filters, offer very high quality factors (about 10,000 at 16 Hz [12] and 2,000 Hz [13]), enabling the design of oscillators with very low phase noise and high stability.

Figure 6.3 shows examples of RF-MEMS.

6.2.2.3   Ultra-Wideband Technology: An Old Dream Coming True [8, 10]

Ultra-wideband was initiated in the early 1900s with Marconi, when spark gap transmitters induced pulsed signals having very wide bandwidths. Spark transmissions created broadband signals, making coordinated spectrum sharing impossible. This type of communications was abandoned for classical narrowband communications. The advent of UWB was enabled by the allocation of Industrial Scientific and Medical (ISM) bands for unlicensed spread spectrum and wideband communications use. Following the Shannon theorem, which states that channel capacity grows linearly with bandwidth but only logarithmically with the SNR, greater capacity could be achieved with large bandwidth allocation. Figure 6.4 shows bandwidth comparisons.

Following this reasoning to its extreme, companies like Time Domain and Xtreme Spectrum started to (re)use pulsed transmissions, hoping that the FCC would allow intentional transmission in power limits; up until then only unintentional emissions were allowed. This was granted, at a level of −41.3 dBm/MHz (known as the Part 15 limit) between 3.1 and 10.6 GHz, providing a

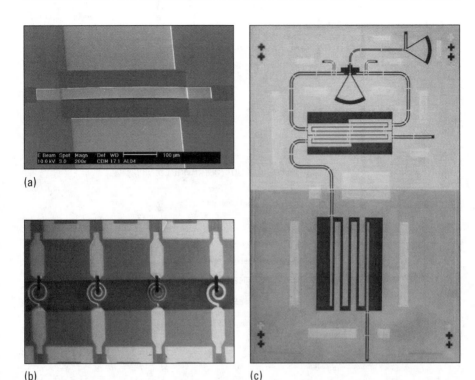

Figure 6.3 (a) Picture of an Aluminum MEMS bridge. The MEMS bridge is 300 μm long, 60 μm wide, with a thickness of 1.0 μm of sputtered Aluminum and is suspened 1.5 μm above the substrate. (b) Micromachined inductors on thin membranes. The resonant frequency 70 GHz for the 1.1 nH inductor, and (c) Micromachined 17–18 GHz single-sideband mixer (filter+balanced mixer). The diodes are still not placed on the wafer. (Courtesy of Prof. Gabriel Rebeiz, University of Michigan, Ann Arbor)

bandwidth of 7.5 GHz (but with an overall power of maximum −3 dBm). (See Figure 6.5.)

As an example, [14] reports capacities of 1 to 18 Mbps at 10m for 1-μW radiated power, which is quite remarkable.

Besides the capacity advantage, UWB has a better immunity to multipath. Unlike conventional wireless technologies, which employ a continuous waveform on a sinusoidal carrier, the UWB signal is pulsed. Individual discrete pulses are essentially unaffected by long delay multipath components (i.e., outside the window of detection). In fact, UWB offers the potential to resolve delays to within a few nanoseconds. As a result, UWB is inherently less susceptible to degradation due to multipath channels. Fading is reduced, enabling lower fade margins to be employed, which saves power. Design complexity and implementation cost can also be reduced, as sophisticated channel estimation, equalization techniques, and interleaving may no longer be required. Improved multipath

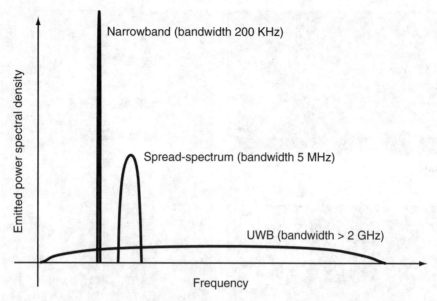

Figure 6.4 Ultra-wideband bandwidth compared to narrowband and spread-spectrum bandwidths.

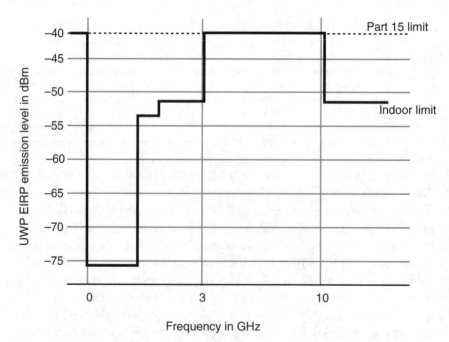

Figure 6.5 ETSI and FCC regulations for UWB.

resistance is a major reason why UWB is of particular interest in support of indoor localization and distance and remote sensing services (i.e., it has potential for resolution of paths to within a few centimeters).

Additionally, UWB has the potential advantage of simplified implementation. It does not require a sinusoidal carrier. For short-range applications, where adequate margin is available to close the link at the required QoS, the intermediate frequency (IF), filters, and amplifiers may be eliminated. Digitally modulated pulses may be transmitted directly over the air interface. Elimination of the RF reduces design complexity and saves cost. It also saves power, since the RF typically consumes a significant amount of the communication power budget. On the other hand, getting rid of these components comes at the cost of a high rate analog to-digital converter, which in turn consumes a lot of power (200 mW for a 3-GHz UWB system). As can be concluded, much research is still taking place.

6.2.2.4 Cross-Layer Optimization

Network design has been greatly eased by a layered approach, where each of the (theoretically seven ISO) layers is designed separately [15]. This approach has proved to work well for wired networks, but it causes large performance losses in wireless networks. The major reason for this resides in the fact that the wireless channel varies a lot over time and space. Fast channel variations of this channel cause a loss in capacity, but little can be done about it. Slow (and/or large-scale) channel variations, on the other hand, can be compensated by adapting the transmission scheme to the channel—basically by using more energy when the channel is good and no energy when the channel is bad. This interaction between the physical and link layer (termed below as adaptive modulation and coding) can be complemented with work on the transport layer (TCP/IP issues) as well as work on the network (multihop ad hoc networking) and application layer (computation/communication trade-offs, adaptive computation).

Adaptive Modulation and Coding

Adaptive modulation and coding [16] involves adapting the modulation, coding, transmit power, and possibly BER, in a fading channel to minimize energy consumption. Typically, in accordance with the QoS required by the application, a BER [or frame error rate (FER)] target is set, and, as a function of the channel (roughly in function of the SNR), the system chooses the best modulation and coding scheme. Hence, unlike for classical mobile systems, one does not need to provide a large fading margin (i.e., transmitting at a higher power than necessary, only to avoid accidental bad communication due to variation in the power loss), which can provide huge energy savings (up to 20 dB).

Ad Hoc Networking: A Path to Low Energy Networks and High Capacity

The first benefit of multihop ad hoc networking can be shown by the following simple reasoning: suppose two nodes are separated by 50m and the propagation law (linking the transmitted energy to the received energy) is given by

$$E_t = E_r d^4$$

where E_t and E_r denote the transmitted and received energy (according to [4], E_r = 0.2 fj for 1 bit); d is the distance between the receiver and the transmitter; and the exponent 4 is typical of indoor situations. A direct link between the two nodes will need 1.25 nj/bit, whereas a 10m link will only need 2 pj/bit; hence, using a multihop ad hoc network would provide about two orders of magnitude gain in energy. (See Figure 6.6.) This opens a huge research domain: energy-aware routing for ad hoc networks, combined with various optimizations at other layers and in particular with wireless transceiver architectures. Indeed, this example is quite naive, as it does not take the total energy consumption of the devices into account.

TCP/IP Issues

The main transport layer protocol for the Internet is the Transmission Control Protocol (TCP). TCP provides two main services:

1. Reliable end-to-end transmission of data, using retransmission of erroneously received packets (as well as reordering of packets to provide a connection-oriented link);

2. Congestion control.

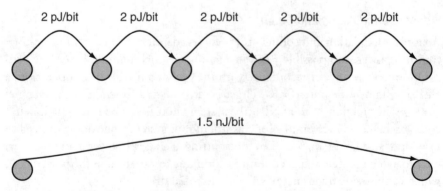

Figure 6.6 Using five hops lowers the necessary energy by a factor of 125.

The first cross-layer optimization that can be provided is between the MAC and TCP layers. Indeed, most wireless MAC layers provide retransmission of erroneous packets, which can consume a lot of overhead time and play the same role as the TCP acknowledgments. By combining the two schemes, the overall delay due to erroneous received packets can be reduced significantly.

The second, and quite famous, cross-layer optimization concerns congestion control. In TCP, packet losses are interpreted as a symptom of network congestion; as a result, the sources adapt the transmission rate with respect to the congestion status. In wireless networks, packet losses are mostly due to bad channel conditions. To alleviate this problem, which is mainly due to changing channel conditions, the lower link layer can use a combination of coding and acknowledgment schemes so that the wireless link is perceived as a constant channel (but with lower capacity).

Adaptive Computation

Adaptive computation involves adapting the computation power used to the required quality of the results. For example, displaying an image on either a 20-in display or on a PDA requires different qualities, and hence computation power. In terms of low power devices, one can also minimize the energy consumed by varying the quality (e.g., of the display) as a function of the remaining power.

Computation and Communication Trade-Offs

Computation and communication trade-offs can be made to waste as little energy as possible, especially in wireless sensor networks. The basic idea is as follows: performing 1 MOP consumes, with current reconfigurable hardware, about 1 μW, which corresponds to the energy that is necessary to transmit 200 bits a distance of 10m (numbers may vary, but it is always possible to compare the computing energy with the transmission energy). The first obvious trade-off involves compression: when the compression effort is lower (in terms of energy) than the number of bits, then compression has to be performed, and vice versa if sending additional bits consumes less than compressing them. Similar arguments also hold for coding.

The second trade-off concerns sensor networks: if an intermediate node has enough data to perform valuable data fusion, it has to trade off the cost of this computation versus the gain in energy (by sending less bits) and decide if data fusion has to be done on the sensor or not.

From our description of energy scavenging to overall optimization of both the network and devices, we can see that the wireless communication landscape changes drastically from an operator- and network-centric approach to a user-centric point of view (for which low power is necessary). Another user-centric characteristic is the wish for true mobile seamless access, which we will now discuss.

6.2.3 The Quest for Mobile Seamless Access

WPANs are part of the quest for mobile seamless access, functioning as a short-range personal access network. Although not presented in the WPAN context, Winlab's Infostation concept is interesting, in that it does not claim the traditional "anywhere, anytime" but rather "many where, many time." Also, it develops an alternative cellular-like type of wireless network to offer high rates in specific places, offering much more than a simple network of hot spots. The mobile seamless access is at the center of most current projects, including the PN-oriented ones detailed below. Terminodes is a good example of this type of project. Its evolution towards a huge lab and a collection of subprojects is also interesting, as it shows that this quest is only attainable with a lot of interdisciplinary work.

6.2.3.1 Infostation

Infostation [17], hosted by Winlab at Rutgers University, is a multistage program aimed at developing network architectures and core technologies for the concept of opportunistic services provided by ultra high-speed short-range radios connected to the Internet. Project aspects include Infostations network architecture, modem/MAC technology, network protocol support (including caching and delayed delivery), and application software. The project also includes proof-of-concept prototyping.

The basic concept is that, instead of having full coverage of a large area with relatively small information rates, the same area is constellated with a number of small hot spots, providing high data rates. The first demonstrated example was that of a highway, on which infostations were present about every 100m. A car passing by the infostation can then, for example, download part of a movie at high speed. When the car goes out of the range of the infostation, it has to wait until it arrives in another infostation's cell, and the fixed network that is linked to the infostations enables the download to continue. This linear array of infostations can, of course, be generalized to a cellular type network for full coverage (but with holes) of a large area.

It is interesting to note that, except for the necessary physical layer–oriented work and the architectural work for the backbone network, most of the research topics that were identified for low-power ad hoc networks are also relevant to this type of approach (i.e., adaptivity and cross-layer optimization).

6.2.3.2 The Terminodes Project (Mobile Information and Communication Systems)

Terminodes [18], a project hosted by Ecole Polytechnique Fédérale de Lausanne, Switzerland (EPFL), proposes to leverage developments in mobile communication and mobile devices in order to give birth to a new kind of mobile information system: a decentralized, self-organizing network based on (mobile) terminals that could simultaneously work as terminals for users as well as network nodes for connecting interuser traffic. These multipurpose terminals are called "terminodes" (terminal + node), and this project aims to popularize the concept of terminodes as the devices used to access services and part of the infrastructure. This project encompasses all layers and explores interlayer interactions, from the fundamentals of the physical layer up to software architecture and applications.

Ubiquity is obviously a central wish of Terminodes, and the solution lies in wireless wide area networks (WWANs) formed by ad hoc networks. Compared to the other projects already presented in this chapter, Terminodes shares most of the cross-layer optimization and high layer challenges, although with a heavier emphasis on software technologies for networks (often fuzzily called middleware). Compared to the projects driven by ambient intelligence and personal networking (detailed below), the initial architectural concept of terminodes, which is less interested in true pervasiveness and embeddedness, leads to quite complex terminals not really suited to the WPAN context.

6.3 Ambient Intelligence: The OXYGEN Project

One of the first and most relevant projects in the field of ambient intelligence is MIT's Oxygen project, devoted to *pervasive human-centered computing*. The project started in 2000, based on initial concepts presented in 1999 in *Scientific American* [19]. The goal of Oxygen is described as "bringing abundant computation and communication, as pervasive and free as air, naturally into people's lives." Oxygen's belief is that computation will be freely available everywhere, like batteries and power sockets, or like oxygen in the air we breathe. This presence will free the user of the need of carrying devices, as configurable generic devices will be present wherever and whenever needed. This belief is the computer equivalent of the telecommunications *anywhere, anytime* dream.

Oxygen identifies several classes of challenges, which are still valid: pervasiveness, adaptation to the environment and user requirements (which implies that Oxygen systems can sense and affect our world), nomadicity, and human/computer interface–related challenges. The Oxygen project is based on the following technologies:

- Device technologies, including (1) computation devices, called Enviro21s (E21s), which are embedded in homes, offices, and cars and can sense and affect the immediate environment, and (2) handheld devices, called Handy21s (H21s), which empower the user to communicate and compute no matter where he or she is;
- Networking technologies, based on so-called N21s;

- Software technologies;
- User technologies (automation, collaboration, and knowledge access/data mining);
- Perceptual technologies.

Oxygen's Networking Technologies

From a strict networking point of view, Oxygen relies mainly on an ad hoc network (possibly mobile) and a gateway to the outer world. The N21s are thus basically MANETs that connect dynamically and can adapt their configurations to the environment and to user's needs. On top of that, N21s integrate other networks (wireless, terrestrial, and satellite) into one seamless *internet*. The core of the N21 is a collaborative region, which is a spontaneous network of computers/devices (which can be part of multiple collaborative regions). The main characteristics of these collaborative regions are as follows:

- Membership to a region is dynamic.
- Resource and location discovery is based on intentional names (for example, a user asks to print on a specific printer). This intent paradigm puts specific constraints on the name resolution and routing schemes as well on the location and service discovery algorithms.
- Security is present at all levels.
- Adaptation is to both the environment (channel condition, networking environment) and the user. This implies the desire for end-to-end QoS (i.e., guaranteeing a specified level of performance on the whole link, the level being a function of the application, the user's intent, and the environment).

6.4   Ambient Intelligence and WPANs: Basis for Personal Networking

Ambient intelligence scenarios developed around the world (from those previously cited to those in the European research community [20]) call for new paradigms in the networking world. Those related to the Internet and core networks are not address here, rather we focus on those that involve the concept of personal networking (the conceptual consequence of the urge for pervasiveness). The three (European) projects described in this section are good examples of the birth of this concept. Indeed, PACWOMAN identified the need for specific WPANs, with rather classic requirements like low power, but also introduced the idea of community area networks (CANs) and the idea of joining different CANs through any wide area network, which is a basic form of a PN.

The concept of PN was introduced by Niemegeers and his colleagues in [21]; it forms the basis for the Dutch national FREEBAND project as well as the large European MAGNET project. Note, however, that the phrase had already been used 10 years earlier [22] and that the concept is also very much linked, at least for the local part, to ambient networking [23, 24].

6.4.1 The PACWOMAN Project

PACWOMAN, which stands for Power Aware Communications for Wireless Optimized Personal Area Network, was initiated based on scenario playing, in which several typical applications were identified. These spanned several orders of magnitude in terms of the required information transfer rate (from 100 bps to 10 Mbps), covering the wireless part of the so-called body area networks (BANs) and extending it to the personal space (up to one's home) and to the outer world (prefiguring the PN concept introduced in the MAGNET and FREEBAND projects).

From a network perspective, PACWOMAN's first goal is to enable the birth of short-range personal networks by using ad hoc networks as self-generating networks. They can be used in sensor networks, small home networks (to exchange information, such as audio/video, alarms, and configuration updates), and as support for information exchange in meetings/conferences.

From a traditional mobile network viewpoint, PACWOMAN opens up a new way of extending mobile networks into the user domain. In this sense, someone accessing a WPAN could use the GPRS/UMTS mobile phone, or a plain old telephone, as a gateway to the Internet or to a corporate IP network.

Taking all these concepts together, the PACWOMAN project is meant to investigate, specify, design, simulate, develop, and demonstrate a true WPAN environment based on IPv6 (and still provide the necessary mechanisms that allow interoperability with IPv4 networks).

To address these different requirements and to enable the design of different classes of devices (from 0.1 Euro to several Euros), the WPAN, according to PACWOMAN, will have to cope with the following requirements:

- *Scalability:* Design an (or a family of) IP radio devices that enable the coexistence of very different data rates (100 bps to 10 Mbps) and a range of different devices (of different complexity/cost);

- *Low-power:* For the low rate segment of WPAN, the multiplicity of very small devices calls for a major breakthrough in battery life (the target will be 1 year). Even for the high rate segment of WPAN, the requirement of autonomy is expected to be very high (at least 1 week), which is a specific constraint that has to be taken into account in the design of both the IP node and the network.

- *Radio integration:* To enable low-cost solutions, the radio, along with the upper layers, has to be integrated on a single chip. Advanced research on mixed signal integration as well as global power optimization will be needed.

- *Wireless nomadic IP:* This is necessary to enable the existence of a wireless IP(v6) radio node, with the power and integration constraints detailed above, and to provide support for ad hoc networking.[3] This includes network management techniques.

- *Security:* This is necessary to insure security on wireless IP nomadic networks and possibly on nonsecure links.

6.4.4.1 Main Objective

The main objective of PACWOMAN is to design a system that will provide a solution to the last-centimeter problem and offer local and global wireless access to the user with bit rates from 100 bps to 10 Mbps with high availability and low power, without requiring infrastructure for local operation. The focus of PACWOMAN is on the global optimization of the device, from the RF to the network layer, and on the global optimization of the network itself. PACWOMAN is also a part of the 4G picture as both a personal and access network, complementing the classic cellular networks and the higher rate/higher cost infrastructure-based WLANs, as well as any other wired/wireless network. Moreover, for global operation, it relies on these wide area networks.

The PACWOMAN system is primarily designed as a network of economical data links:

- Between devices linked to a person;
- Between two or more persons in vicinity (1 to 5m) of each other;
- Between persons and their environment;
- Between persons and the outer world (via third-party wired or wireless networks).

In order to build a comprehensive solution for the end user, the three lower OSI layers of PACWOMAN are evaluated and developed within this project, along with security mechanisms.

Since PACWOMAN is a flexible wireless system offering a large range of distinct data rates and leading to a variety of devices, it offers definite advantages over Bluetooth or 802.15:

- The enhanced cooperation between very different devices (without resorting to duplicate hardware);

- The optimized ad hoc networking capabilities (including ad hoc network management and optimization);
- The ability to link cost and power to the useful data rate;
- The ability to design very low cost devices for high mass markets (like RF-tags).

6.4.4.2   The WPAN According to PACWOMAN

WPANs are expected to be a major part of the so-called 4G picture. Indeed, WPANs will provide the user with true seamless data and multimedia communication between him and his family, friends, and colleagues, and with the help of the expected *exploded terminal*, he will not have to carry heavy terminals nor need to charge their batteries again and again. PACWOMAN is based on a user-centric concept and has defined its main characteristics and challenges, which are extreme low power, scalability in user rates and in type of devices, ad hoc networking, and security challenges. To meet all these requirements, a novel design approach is needed, where the main novelties lay in the codesign of nearly all ISO layers with security and in the global optimization of the communication node and link.

To introduce a really user-centric WPAN concept, and to make technical choices that serve the users' purposes, the first step was to develop key user scenarios and derive preliminary communication needs from these scenarios. Based on three typical situations (the smart healthy home, the professional environment, and the fancy futuristic multimedia traveler), the conclusion was drawn that scalability and end-to-end networking with security were the most important concerns, with the additional constraints that the system could comprise a lot of devices and that devices should operate without battery charging for at least 6 months. The next step was to further develop the layered network architecture by developing ideas about the influence of the scenarios and network architecture on the physical layer(s) and devices. Some shortcomings of current systems are put to light through the characteristics of the proposed WPAN system.

6.4.4.3   The PACWOMAN Vision: On devices and Data Rates

The first lesson learned from the envisioned scenarios is that a lot of different data rates and terminal functionalities are used. To address this wide range of data rates, two basic options are possible:

1. Different physical layers (2 or 3), where each addresses a data rate range (e.g., 10 bps to 10 Kbps, and 10 Kbps to 10 Mbps);
2. Scalable physical layers (data rates, power, and cost).

Clearly, certain devices will be more capable and costly than others. Simple personal devices (e.g., sensors) must be very low cost, and certain less capable devices may even be of the throwaway variety. Other more capable devices may incorporate bridge, router, or even gateway functionalities, as required to support advanced networking features. Nonetheless, the additional cost of adding WPAN functionalities to more capable (and secured) personal devices (e.g., UMTS handset) should not be more than a small fraction of the cost of the basic device. Relative to other wireless technologies, the PACWOMAN approach should be inherently low cost, due to a scalable and hierarchical architecture and to (possibly multiple) air interface options tailored to the service class.

6.4.4.4 The PACWOMAN Vision: A Three-Level Network

A second lesson learned from the scenarios is that communications occur in three different spaces: the space that is centered on the person itself, the outer local space, and the outer distant space. These three spaces translate themselves into three possible networks: the personal area network, the community area network, and the wide area network. The PAN will address the whole range of data rates, with possibly many very small devices (supporting very low data rates) and a small number of more-capable devices like PDAs, displays, and cameras. Hence, the PAN in itself can be considered as having two classes of devices (along with two classes of communications). Also, it seems quite natural to divide the PAN itself in two networks: one that is formed by the low data rate devices and the other formed by of the medium to high rated devices. In this description, we focus mostly on the first two levels (PAN and CAN), and only give some hints about the WAN issues.

First Level: The Personal Area Network

In the simplest case, the PAN may be a stand-alone network capable of independent operation. Still, due to the very large range of data rates, it is useful to inject a hierarchy into this simple network by separating the low rate devices from the high rate ones.

The Low Rate PAN: A Virtual Device. A virtual device (VD) is made of two types of devices: basic terminals (bTs), which can be very simple telemonitoring sensors or actuators, and a master (M). Indeed, as the distance between these devices is short (about 2m for a person) and as the use of direct communication between the BTs does not appear to be mandatory, the natural topology is a star topology. Hence, there needs to be a master that coordinates the communications and can serve as a display/control terminal. The whole network acts as a concentrator from low data rate to higher data rate through the master, and it can be seen as a virtual device from the high rate PAN network (Figure 6.7).

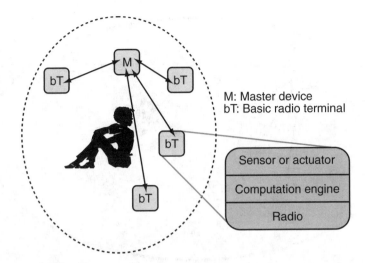

Figure 6.7 The virtual device: a small standalone network.

The High Rate PAN: A Personal Area Network. Apart from small sensors, a person could also carry a camera, a display (e.g., virtual glasses on which one could see an electronic display), or even a more computer-oriented device (e.g., computer, keyboard, or printer). To accommodate a higher data rate without wasting bandwidth by duplicating communications (which would occur if the data going from the camera to the display has to go through the master), the natural topology would be a meshed network (hopefully fully connected, but this cannot be guaranteed). Hence, the global picture of a PAN is a meshed network where one of the nodes is a VD, which groups the low data rate devices, and the other nodes are advanced terminals (aT) (Figure 6.8). Note that dynamic reconfiguration of the network and security issues are less crucial and complex at this level than for the CAN or WAN.

Second Level: The Community Area Network

The novelty of the PACWOMAN concept is focused on its ability to offer advanced networking functionalities and information services. Specifically, the concept includes ad hoc networking that enables the formation of networks anytime, anywhere, while maintaining the integrity of information and applications within an individual personal area space. To support ad hoc networking, the network must be made of bridges (B) (devices that can handle Layer 2, in the OSI taxonomy) and routers (R) (devices that can handle Layer 3). The network itself will be a meshed network and, to enable compatibility with most other networks, will be a packet-based IP network. Network elements may be static or mobile, and CANs may be formed between two or more PANs, or equivalent network entities. Figure 6.9 provides an illustration.

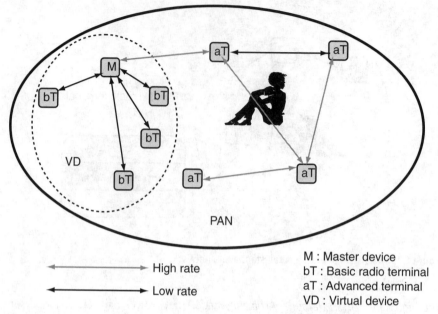

Figure 6.8 The personal area network: a network of terminals.

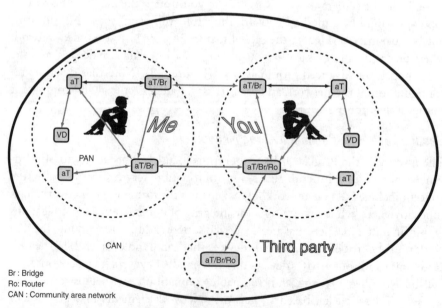

Figure 6.9 The community area network: a local network of PANs.

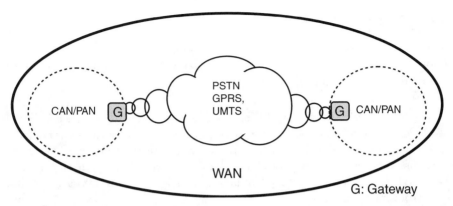

Figure 6.10 The wide area network: PANs and CANs communicating through external networks.

In this concept, the user's PAN becomes a private, virtual domain. The capabilities of the user's PAN may be tailored to support a range of services and QoS requirements and may be scaled in accordance with the available resources per service class and adapted to the prevailing networking capabilities. Tailoring PAN capabilities includes the adaptation to available resources and downloadable software radio interfaces (i.e., for more capable devices), as well as the enhancement of a basic handset or PDA with new software or services (which means that reconfigurable hardware and hardware/software partitioning will be necessary).

Gate-Keeping Functionality. A person will certainly want to control his or her privacy and thus will need a gatekeeper to take care of that. This gate-keeping functionality can be tackled by security mechanisms that already exist in Internet networks, for example. Still, the gate-keeping functionality could be concentrated on one device, or totally distributed among the terminals. Many of these mechanisms could be borrowed from existing security schemes, but research will continue to focus on finding low complexity solutions still with a very good level of security (hopefully better than on the Internet).

Third Level: The Wide Area Network

To complete the picture, the system must provide global communication possibilities to the user, which calls for the use of classical WAN systems (wireless or not). To enable this, communications over these systems go through a gateway (Figure 6.10). Issues that arise here are end-to-end QoS and security.

6.4.2   The MAGNET Project

The European MAGNET project, which stands for My Personal Adaptive Global Network, goes a step further than PACWOMAN, as it incorporates the personal networking concept [25]:

> The MAGNET vision is that Personal Networks (PNs) will support the user's professional and private activities without jeopardizing privacy and security. This support will take place through the user's own personal network (PN) consisting of a core Personal Area Network (PAN) extended with clusters of remote devices which could be private, shared, or public and able to adapt to the quality of the network accessed.

To realize this vision, MAGNET shares the following objectives with previously cited projects:

- PN architecture;
- Security and privacy issues in PNs;
- Adaptive and (re)configurable radio access covering a wide range of data rates, system capabilities, and requirements, optimized for low power and cost-effectiveness;
- Networking and interworking issues both at the PN and PAN level, particularly resource and context discovery, self-organization, mobility management, addressing and routing, service discovery, and cooperation between public and private and licensed and unlicensed networks.

This last point is much more important in the PN concept than in that of a WPAN. Interworking and heterogeneous networking is a major challenge and research topic within MAGNET.

On top of the technical issues presented above, MAGNET places emphasis on building insight into business models for PNs and the related mobile data services in multinetwork environments and has embraced nontechnical partners to help develop a truly user-centric concept (like what was done for the EU ambient intelligence scenarios and the WWRF Book of Visions [26, 27]).

PNs comprise potentially "all of a person's devices capable of network connection whether in his or her wireless vicinity, at home or in the office" [25] (Figure 6.11). Hence, a combination of the PAN and ambient intelligence (AI) paradigms has to take place, as well as an adaptation and extension of both concepts. The extension of the PAN concept (which is strictly speaking a 10m range network) is that PNs have an unrestricted geographical span.

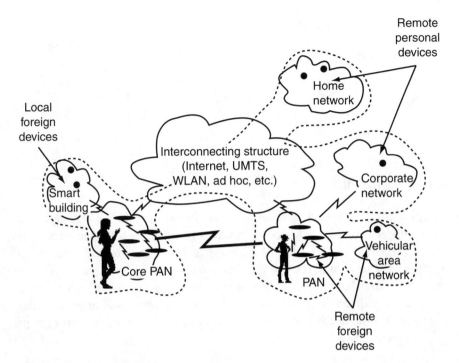

Figure 6.11 Personal network.

Pretty much like PACWOMAN's CAN, it may incorporate remote devices, whatever their geographic location, resorting to either ad hoc networking or infrastructure-based networks (up to the Internet). The adaptation and extension of the AI paradigm consists in the ad hoc and automatic configuration of the networking architecture itself, borrowing all the user-centric and environment awareness of AI.

To conclude, the main components of a PN are [25]:

- A core PAN consisting of personal devices in the close physical vicinity of a user, including devices moving along with him or her. The core PAN is an essential component of the PN.

- Local foreign devices or clusters thereof, which are owned by other parties and could either be reserved solely for the PN owner or be shared with others. They are linked to the core-PAN via communication infrastructures.

- Remote personal devices, which are grouped into cooperating clusters and which are linked to the core-PAN via communication infrastructures.

- Remote foreign devices or clusters thereof, which are linked via communication infrastructures and can be shared with many users or be reserved for the PN owner.

- Communication infrastructures, in principle WANs making use of some sort of infrastructure-based resources (CN), which can be public (e.g., cellular, Internet) or private (e.g., leased lines), licensed or unlicensed (e.g., WLAN).

6.5 Conclusions

Escaping for a moment from technical considerations, the authors and readers dream of a world where people count! The (industrial and academic) research projects presented here are meant to make this dream true, even if intellectual and commercial considerations are at stake in these projects. At the very least, these developments, applied to the scenarios envisioned in Chapter 5, will enable people to benefit from technological advances relating to their basic needs (e.g., health), their business, or merely their comfort (e.g., entertainment).

References

[1]   Weiser, M., "Hot Topics—Ubiquitous Computing," *IEEE Computer*, Vol. 26, No. 10, October 1993, pp. 71–72.

[2]   ITEA Ambience Project, http:www.hitech-projects.com/euprojects/ambience.

[3]   Rabaey, J. M., et al., "PicoRadio Supports Ad Hoc Ultra-Low Power Wireless Networking," *IEEE Computer Magazine*, Vol. 33, No. 7, July 2000, pp. 42–148.

[4]   Rabaey, J. M., "Ultra Low-Power Computation and Communication Enables Ambient Intelligence," *Smart Objects Conference*, Grenoble, France, May 15–17, 2003.

[5]   Meninger, S., et al., "Vibration-to-Electric Energy Conversion," *IEEE Transactions on VLSI Systems*, Vol. 9, No. 1, February 2001, pp. 64–76.

[6]   Sinha, A., A. Wang, and A. P. Chandrakasan, "Energy Scalable System Design," *IEEE Transactions on VLSI Systems*, Vol. 10, No. 2, April 2002, pp. 135–145.

[7]   Min, R., et al., "Energy-Centric Enabling Technologies for Wireless Sensor Networks," *IEEE Wireless Communications*, Vol. 9, No. 4, August 2002, pp. 28–39.

[8]   Warneke, B., et al., "Smart Dust: Communicating with a CubicMillimeter Computer," *IEEE Computer Magazine*, January 2001, pp. 2–9.

[9]   Roundy, S., et al., "A 1.9GHz RF Transmit Beacon Using Environmentally Scavenged Energy," *IEEE Int. Symposium on Low Power Elec. and Devices*, 2003.

[10]   Barrett, T. W., "History of UltraWideBand Communications and Radar: Part I, UWB Communications," *Microwave Journal*, Euro-Global Edition, Vol. 44, No. 1, January 2001, pp. 22–56.

[11]   Roundy, S., P. K. Wright, and J. M. Rabaey, *Energy Scavenging for Wireless Sensor Networks with Special Focus on Vibrations*, Boston, MA; Kluwer, 2003.

[12]   Maxey, C. A., "Switched-Tank VCO Designs and Single Crystal Silicon Contour-Mode Disk Resonators for Use in Multiband Radio Frequency Sources," M.A. Thesis, Virginia Polytecnic Institute and State University, 2004.

[13]   Mukherjee, T., et al., "Reconfigurable MEMS-Enabled RF Circuits for Spectrum Sensing," *Government Microcircuit Applications and Critical Technology Conference 2005 (GO-MACTech '05)*, Las Vegas, April 4–7, 2005.

[14]   Le Boudec, J.-Y., et al., "A MAC Protocol for UWB Very Low Power Mobile Ad-Hoc Networks Based on Dynamic Channel Coding with Interference Mitigation," *Technical Reports in Computer and Communication Sciences*, EPFL, 2004.

[15]   Shakkottai, S., T. S. Rappaport, and P. C. Karlsson, "Cross-Layer Design for Wireless Networks," *IEEE Communications Magazine*, Vol. 41, No. 10, October 2003, pp. 74–80.

[16]   Chung, S. T., and A. J. Goldsmith, "Degrees of Freedom in Adaptive Modulation: A Unified View,"*IEEE Transactions on Communications*, Vol. 49, No. 9, September 2001.

[17]   Frenkiel, R. H., et al., "The Infostations Challenge: Balancing Cost and Ubiquity in Delivering Wireless Data," *Personal Communications*, IEEE, Vol. 7, No. 2, April 2000, pp. 66–71.

[18]   Hubaux, J.-P., et al., "Toward Self-Organized Mobile Ad Hoc Networks: The Terminodes Project," *Communications Magazine*, IEEE, Vol. 39, No. 1, January 2001, pp. 118–124.

[19]   Dertouzos, M., "The Future of Computing," *Scientific American*, August 1999.

[20]   Raychaudhuri, D., "ORBIT: Open-Access Research Testbed for Next-Generation Wireless Networks," proposal submitted to NSF Network Research Testbeds Program, May 2003.

[21]   Niemegeers, I. G., and S. M. Heemstra de Groot, "From Personal Area Networks to Personal Networks: A User Oriented Approach," *Wireless Personal Communications*, Vol. 22, No. 2, August 2002, pp. 175–186.

[22]   Braun, K., et al., "Universal Personal Networking," *Universal Personal Communications, 1993, Personal Communications: Gateway to the 21st Century, Conference Record*, Vol. 1, October 12–15 1993, pp. 108–112.

[23]   Duda, A., "Ambient Networking," *Smart Objects Conference*, Grenoble, France, May 15–17, 2003.

[24]   Niebert, N., "WWI Ambient Networks Project Overview and Dissemination Plan," June 14, 2004.

[25]   Prasad, R., "MAGNET: My Personal Adaptive Global Network," *IST-Mobile Summit*, Lyon, France, June 28–30 2004.

[26]   Ducatel, K., et al., (Eds.), "Scenarios for Ambient Intelligence in 2010," IST Advisory Group (ISTAG), European Commission, Brussels, 2001.

[27]   Mohr, W., et al., (Eds.), "The Book of Visions 2001," (draft) Version 1.0, Wireless Strate-
       gic Initiative, December 2001.

[28]   "Adoptive Multi-domain Power aware Sensors," http://www.mtc.mit.edu/research-
       groups/icsystems/vamps.

Selected Bibliography

"Composite Reconfigurable Wireless Networks: The EU R&D Path Toward 4G," *IEEE Communications Magazine* topical issue, Vol. 41, No. 7, July 2003.

"Context Aware Pervasive Computing," *IEEE Wireless Communications* topical issue, Vol. 9, No. 5, October 2002.

Mandke, K., et al., "The Evolution of Ultra Wide Band Radio for Wireless Personal Area Networks," *High Frequency Electronics*, Vol. 2, No. 5, pp. 22-32.

Win, M. Z., and R. A. Scholtz, "Impulse Radio: How it Works," *IEEE Communications Letters*, Vol. 2, No. 2, February 1998, pp. 36–38.

7

Mobile Ad Hoc Networks (MANET)

MANET's roots lay in the first packet radio networks (initiated in the 1970s at the University of Hawaii) and in the Hams, the amateur packet radio networks that flourished in the 1980s. In the early 1990s, mainly triggered by work on WLANs, significant contributions were made on the MAC layer. At the end of the 1990s, the Internet Engineering Task Force (IETF) created the MANET group [1, 2], which works mainly on the routing aspects of mobile ad hoc networks and has now evolved to a rather mature collection of routing algorithms. Rather recently, the introduction of multiple antennas and cooperative diversity opened new research possibilities, promising significant enhancements in the network capacity as well as in the energy efficiency of MANETs. In the rest of this chapter, we provide a short introduction to MANETs and its link with WPANs, as well as an overview of the MAC and routing aspects of MANETs. We follow with a description of more recent work on network capacity, along with an introduction to multiple antennas and cooperative diversity.

7.1 Introduction

The MANET working group of the IETF has given mobile ad hoc networks the following description [2]:

> A "mobile ad hoc network" (MANET) is an autonomous system of mobile routers (and associated hosts) connected by wireless links—the union of which form an arbitrary graph. The routers are free to move randomly and organize themselves arbitrarily; thus, the network's wireless topology may change rapidly and unpredictably.

Focusing on WPANs, we show that MANETs are an important concept for personal area networks, and hence for all applications that are targeted by WPANs.

7.1.1 Why MANET in WPANs?

Originally designed as a simple cable replacement, WPANs are evolving towards complex networks of various mobile devices with very different capabilities, like those shown in Figure 7.1 (reprinted from Figure 6.9).

Although simplistic, this WPAN architecture is typical of what IETF defines as a mobile ad hoc network, as the global WPAN system acts as an autonomous system of mobile routers, which are free to move and capable of organizing themselves. Hence, WPANs could well be the *killer application* for ad hoc networks.

7.1.2 Historical Overview

The initial step towards MANET was the ALOHA project [3], which used the broadcast nature of radios to design a single-hop data radio network. ALOHA was followed by a multihop project, named the Packet Radio Network (PRNET) [4], whose architecture is quite close to the current view of a MANET. Indeed,

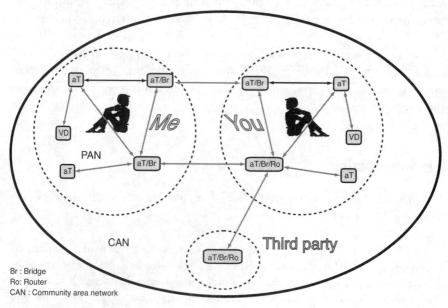

Figure 7.1 A typical WPAN.

a PRNET comprises mobile terminals and mobile repeaters (prefiguring the mobile routers), which can forward packets from any source to any destination. Routing is based either on:

- *Point-to-point routing*, with each packet traveling from source to destination through a predetermined sequence of repeaters; whereas the sequence (route) is determined by a mobile station that has a global knowledge of the network. This routing strategy works well for slowly varying network topology.

- *Broadcast routing*, where the packets are then flooded to the whole network, introducing a robust (with respect to network topology change) but greedy (hence power inefficient) routing protocol.

A major contribution of PRNET was the introduction of the CSMA algorithm. CSMA is a simple protocol that prevents a device from transmitting a packet when a neighboring device is using the medium. This is done by simply sensing the medium, and when the medium is found to be busy (i.e., a neighboring device is transmitting a packet), the device will stay quiet.

During the 1990s, a number of projects that were inspired by PRNET led to the development of ad hoc routing algorithms, and eventually led to the creation of the IETF MANET group [1, 2]. This group focused mainly on routing algorithms with various goals (connectivity, throughput, energy conservation) but evolved to a broader research scope (including advanced MAC layers, network capacity, multiple antennas, and cooperative diversity).

7.1.3 Main Features

A mobile ad hoc network is an autonomous system of mobile routers (and associated hosts) connected by wireless links, the union of which forms an arbitrary graph. The routers are free to move randomly and organize themselves arbitrarily; thus, the network's wireless topology may change rapidly and unpredictably.

The RFC 2501 by MANET working group of the IETF [2] points out the following features as some of the most relevant characteristics of MANETs.

7.1.3.1 Wireless Related Features

The network capacity can be very low, as the throughput for each node is much less than the radio's maximum data transmission rate. Indeed, this throughput is limited by the wireless channel's impairments, such as noise, fading (including multipath and Doppler effects), interference from other sources (among other multiple access interference), the limited availability of spectrum, as well as by the inherent problems the medium access control protocol has to deal with.

Hence, the capacity of the wireless links is always much lower than in the wired and optical counterparts. Moreover, Kumar and Gupta [5] showed that the throughput available per node in an ad hoc network decreases with the size of the network, indicating that this throughput is much lower then the data rate available at the link level. Even if mobility and/or architectural work can alleviate this capacity loss, the gap between data rate and effective throughput (sometimes called *goodput*) will remain important.

Spectral reuse, on the other hand, can be quite high. Indeed, using short-range communication links, as in WPAN and MANET, the emitted power can be kept very low, introducing very low interference levels and allowing for the use of a high spectral reuse.

Moreover, very high data rates (up to 1 Gbps, as introduced by IEEE 802.15) make it possible to build (at this moment small but true) multimedia "mobile" ad hoc networks (as long as the nodes are less than a couple of meters apart).

Robustness, challenged by the wireless channel and mobility, will remain an issue and necessitate a lot of work at all layers, especially at the network layer, as will be shown in the next section.

7.1.3.2 Dynamic Topologies

Nodes can move freely in arbitrary directions and speeds. Therefore, the network must adapt itself to unpredictable changes in its topology, which is typically multihop.

Fast installation and mobility are the main desirable features of MANET, since setting up a network does not require any preliminary infrastructure and is done automatically, without user interaction (at least if the user's profile allows it). Mobility of the nodes is handled by the routing algorithms, so that the user can move without being aware of the networking aspects, as long as the nodes do not move too fast and the connectivity of the network is ensured (i.e., the network is not partitioned in two distinct subnetworks that do not see each other). Note that the equivalent of cellular mobility management concepts such as roaming and location areas are absent here, as mobility management is included in the routing algorithms.

Fault tolerance is also an important feature (and challenge) of MANET, as MANET has to support connection failures due to mobility. Hence, fault tolerance must be handled by all MANET layers (from link layer to the application, but most of the effort has been done at the routing and transport layers).

Localization of the nodes is important, as it can help in designing efficient routing algorithms (see Section 7.3). Localization is also important at the application level, as it can offer location-specific services to the user, which are an inherent part of the personal networks introduced in Chapter 6.

The scalability of MANETs in terms of number of nodes, though less an issue for WPANs, which are not expected to be huge networks, is important for application fields like sensor networks. Indeed, in this case, the number of nodes can be quite high (see, for example, projects like SmartDust in Chapter 6). As MANET research has been partly funded by the U.S. Department of Defense, these aspects have been largely handled.

Scalability in terms of type of nodes (i.e., more capable nodes, which have less energy constraints versus dumb nodes, which are self-powered and have limited capabilities) is also an important issue, but one that received less attention in the early days of MANET and is mainly linked to the energy-constrained operation.

7.1.3.3 Energy-Constrained Operation.

The devices that form part of the WPAN will be power limited, as they will be (mostly) portable and possibly self-powered (as in sensors networks, for instance, where maximizing the average network life is a design criteria). The main issues related to the energy constraints are:

- *Routing algorithms:* According to the design criteria (individual power saving, average network life), different variations of routing algorithms will be designed; they will be driven by the device's processing capability.

- *Processing capability:* Although current WPANs are mainly linked to mobile phones and automotive applications, sensor network applications and in general the low-rate WPAN devices will have very low processing capabilities, and hence will probably not be able to play routing/relay roles. The design of the WPAN/MANET architecture will be pretty much application dependent. In the scenario where a MANET is be based on the Bluetooth part of mobile phones, the energy used for routing/relaying information has to be minimized, since the user would refuse to loose battery life for it to serve as a router.

- *Battery life:* This is one of the main design criteria for routing and power control algorithms (see, for example, [6, 7]).

- *High latency:* This is an unwanted by-product of most energy-conserving designs. These designs usually put the nodes in sleep or idle mode, in which the node cannot forward packets, thus slowing the routing process and increasing latency.

7.1.3.4 Limited Physical Security

Recent security attacks against WLANs and Bluetooth-enabled mobile phones have illustrated the vulnerability of wireless networks. Existing security techniques at the link layer and network layer can be applied quite easily (and can provide a good level of security) in WPANs and MANETs, but they are subject to the following caveats [8]:

- To achieve scalability in terms of device capability, different levels of security should be provided, to enable simple and low power mechanisms for the low rate, low power devices.
- To achieve scalability in terms of number of nodes, a clear structure of the whole security architecture should be designed, in order to ensure end-to-end security.

Besides these specific caveats (which have a big impact on the security architectural designs), the classic approach to security in wireless networks can be applied for WPAN/MANETs (this will be detailed in Chapter 8).

7.2   Medium Access Control

Medium access protocols are responsible for the sharing of the common (wireless) channel. The main objectives of MAC layer protocols are:

- Throughput maximization;
- Delay minimization (access delay plus transmission delay);
- Fairness of access;
- Geographic scalability (i.e., capability of being applied in large networks);
- Rate/QoS scalability (i.e., capability of being applied to low rate and high rate networks, as well as networks where different type of services are offered);
- Robustness against interference;
- Energy efficiency.

MAC protocols [9] can be divided into three categories (see also, Figure 7.2) (other categorizations can be found in the literature [10]):

1. Fixed assignment like TDMA, FDMA, and CDMA, which are used in cellular systems (since they are rarely envisaged for MANETs, we will not discuss them here);
2. Random assignment including protocols like (slotted) ALOHA and CSMA;

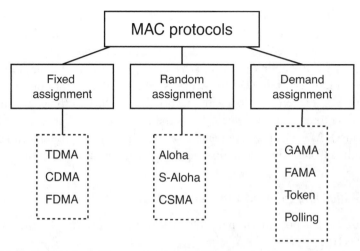

Figure 7.2 A classification of the most common MAC protocols.

3. Demand assignment, which tries to combine the best of the previous two methods. This class includes protocols like Token Ring, FAMA, GAMA, and polling schemes.

7.2.1 Common Random Assignment Protocols

The first protocol is ALOHA, name after the Hawaiian word "hello"[1] and developed in the first radio packet network at the university of Hawaii. In this system, when a packet arrives at the radio, it simply says hello and the packet is sent. Upon correct reception, the receiver sends an acknowledgment; otherwise, after a random time, the source sends the packet again. ALOHA offers a maximum throughput that is 18% of the maximum data rate and is hence quite inefficient. Inefficiency comes from the high rate of collisions between packets. In particular, when a packet is in the air, even if a large portion of it has been transmitted without problem, another packet can be sent at that moment, resulting in a collision and the loss of nearly two packet durations. Slotted-ALOHA (S-ALOHA) solves this problem by ensuring that all packets are synchronized; hence, the collision always takes place at the beginning and the efficiency can go up to 36%. S-ALOHA is used in the reservation phase of mobile networks like GSM and UMTS.

---

1. The Hawaiian word "Aloha" is actually a general greeting. It means "hello," as well as "good bye." A friendly and helpful person is said to have lots of Aloha.

The other major contention-based protocol (i.e., where different users contend for the channel) is CSMA. In CSMA, the user listens to the channel before sending. If the channel is idle, it sends it packet either directly (in 1-persistent CSMA), with a probability p (in p-persistent CSMA), or after a randomly chosen time (in nonpersistent CSMA, which is used in 802.11, for example, where the time is larger when collisions have occurred). In CSMA, efficiency can go up to 80% for rather small networks. For large networks (i.e., large distances between nodes), ALOHA is still preferable, as its efficiency does not depend on distances unlike CSMA, whose performance degrades with the size of the network.

7.2.2 Hidden and Exposed Terminal Problems

The hidden and exposed terminal problems are well known in contention-based protocols [10].

The hidden terminal problem occurs when two transmitters are not in the transmission range of each other (i.e., they cannot receive each other's signal) but are in range of a common receiver; therefore, carrier sensing fails in this case as both transmitters will sense a clear channel but will cause a collision at the receiver.

The RTS/CTS scheme can solve the hidden terminal problem, as illustrated in Figure 7.3. In this figure, nodes A and C are hidden from each other. When A wants to transmit data to B, it asks for permission by sending a RTS, to which B answers with CTS. As C is within transmission range of B, it overhears the CTS and does not transmit its packet (i.e., in 802.11, CTS carries the duration of the data packets, hence C knows how long it has to remain silent).

In the exposed terminal problem illustrated in Figure 7.4, terminal C will not send data to D, although it could, leading to inefficient use of the channel. Indeed, if the RTS/CTS scheme is not used and if node B transmits data to node A, node C hears the communication and refrains from communicating. Looking at Figure 7.4, one can see that the receiver A is not in the range of C, hence C

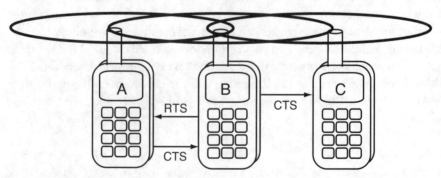

Figure 7.3 RTS/CTS scheme to solve the hidden terminal problem.

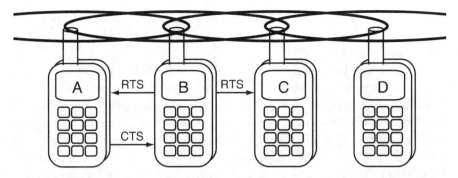

Figure 7.4 The exposed terminal problem can be worsened by the RTS/CTS scheme.

could in fact send data to D without collision. Considering the RTS/CTS scheme, C overhears the RTS but does not receive the CTS, hence C knows that B is sending data to A and knows that A is not in his range, hence C can send data to D, which partly solves the exposed terminal problem. Other solutions to the exposed node terminal consist in using separate control and data channel or to use directional antennas, as will be discussed below.

7.2.3 MACA, MACA-BI, MACAW (for Wireless LANs), and 802.11

Multiple access collision avoidance (MACA) [11] was the first wireless protocol to apply RTS/CTS handshaking. MACA basically uses a three-way handshake (RTS-CTS-Data) without using carrier sense. Moreover, MACA uses the knowledge of the power with which the RTS/CTS are received to implement power control (i.e., sending no more power than needed to reach a specific node). Collision avoidance (reduction of the time in which the channel is not used due to the collision) comes from the fact that collisions mainly occur in the RTS/CTS phase, in which the control packets have a much smaller duration than data packets; hence, the probability of collision is smaller and the duration of each collision as well. By drastically lowering the collision time and by transmitting data in a contention-free period, MACA is often considered as a collision-free protocol.

MACA By Invitation (MACA-BI) avoids half of the control packets by imposing the receiver to send an invitation (the CTS, which is renamed the RTR—request to receive). The main disadvantage of MACA-BI is the implicit assumption that the receiver knows when the emitter has a packet to transmit. Hence, for bursty communications, MACA-BI is rather inefficient. On the other hand, for constant bit rate communications, the receiver can easily determine when to send RTRs.

MACAW [12] appears as a rather natural extension to MACA and combines CSMA with MACA, with the addition of an acknowledgment to ensure more robustness.

802.11 uses the Digital Foundation Wireless MAC (DFWMAC) protocol, which is an extension of MACAW, using a CSMA/CA protocol with an optional RTC/CTS scheme. In DFWMAC, the transmitter senses the channel: if the channel is idle, it sends its packet directly; otherwise, it chooses a random number cw between 0 and CW-1 (CW means contention window) and tries again after ($cw * a\_slot\_time$) seconds. If the transmission is not successful (the channel is not idle), it doubles CW and chooses again a random number of slot times to wait; otherwise, it sends its packet. After a successful transmission, the algorithm starts again by sending a packet directly upon channel idleness.

One of the main drawbacks of the 802.11 MAC protocol is the so-called *silenced zone*, which is shown in Figure 7.5. In this figure, a transmitter (T) requests a communication with a receiver (R). By the exchange of RTS/CTS, and according to the power sent by the T/R pair, the nodes that are in the transmission range for T and R have to keep silent during the whole transaction, and hence their access to the medium is denied. This silenced zone has to be kept at a minimum to increase the spatial reuse (indeed, if one divides the transmission

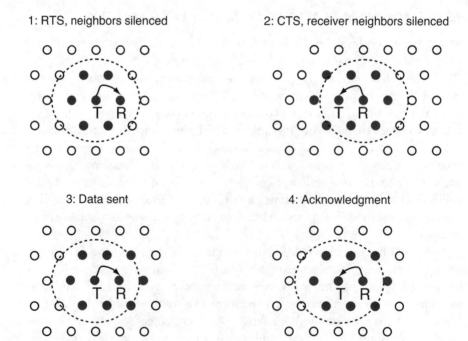

Figure 7.5 The silenced zone created by the 802.11 MAC.

range by two, the number of hops to go from one place to the other is multiplied by two, but the silenced zone, which is proportional to the square of the transmission range, is divided by four, yielding a larger network capacity). The impact of the silenced zone is so large on the capacity that many MAC protocols try to alleviate the capacity problem incurred (see the directional/multiple antenna MAC as well as power control algorithms discussed below).

7.2.4 Demand Assignment MAC Algorithm

Floor acquisition multiple access (FAMA) [13] represents a family of MAC protocols, which operate in two phases: first the channel, or floor, is acquired and then the actual data is transmitted. Once the channel is obtained, the transmission is guaranteed error-free.

The fundamental method for ensuring that only one user captures the channel is the application of a CTS message with a longer duration than the RTS message. The length of the CTS is made longer than the length of a RTS (i.e., it is increased by the maximum round-trip time, the turnaround time Tx/Rx, and some processing time). This ensures dominance of the CTS over the RTS: when a RTS and a CTS message collide, the CTS message can be detected after the collision, thus reserving the channel so that transmission belonging to the session from which the CTS originated can be continued successfully.

The FAMA protocol, however, has not been designed to provide QoS; if the control packets collide, the retransmissions are scheduled randomly making it difficult to impose delay constraints on sessions.

Several variants of the FAMA family exist; the aforementioned version is referred to as nonpersistent carrier sensing (NCS). Another variant is nonpersistent packet sensing (NPS). In these protocols the stations do not sense the channel before transmitting. The basic working of this protocol is similar to that of MACA; however, for these protocols to work with hidden terminals, the CTS messages have to be transmitted multiple times. It was shown that these carrier-sensing protocols are much more efficient than the traditional protocols such as MACA.

Group allocation multiple access with packet sensing (GAMA-PS) [14] divides the channel into cycles, where there is first a contention period to obtain membership of the group of users that is allowed to transmit over the channel, and the second is a collision-free transmission period for the members of this group. This means that some kind of scheduling of the transmission in this period is required.

This protocol includes aspects from both CSMA and TDMA. In the lightly loaded case the number of users in the transmission group is low and the system behaves like CSMA, whereas in the highly loaded case many users are members of the transmission group and the system behaves more like a TDMA system, as (almost) all transmissions are scheduled.

These protocols deviate from "pure" CSMA in the sense that, based on the reservation request (to enter the transmission group), each node constructs the channel state information (resulting in a more scheduled than random approach).

Other on-demand MAC protocols have been proposed, all based on some form of reservation system, and they are mainly extensions of or corrections to FAMA and GAMA.

7.2.5 Multichannel MAC Protocols

Busy tone multiple access [15] can be considered the simplest multichannel protocol, as it assumes the presence of two channels: a data channel and a control channel. While receiving data, a busy tone is transmitted on the control channel. This signal informs other users that they should defer their transmissions. To further increase the performance of wireless networks, multiple data channels can be used, as in CDMA, for example.

The main advantage of multichannel MAC is that QoS can be supported more easily [16, 17], since resorting to multiple channels introduces a higher level of granularity in the sharing of the bandwidth. Hence, each channel can be considered as a quantum of data rate, and providing a given data rate (which is one QoS parameter) simply translates to granting a given number of channels.

Multichannel CSMA [18] is a straightforward extension of CSMA where the carrier sensing is done on n channels rather then on one. The main drawback is the hardware complexity introduced by this protocol, as each device has to monitor n channels in parallel. Suppose that there are n channels; this protocol applies the traditional CSMA, only now it monitors if one of the n channels is available. If so, the data is transmitted on the available channel. This protocol does not solve the hidden terminal problem and has high hardware costs.

Another proposed protocol is the hop-reservation protocol based on slow frequency hopping spread spectrum (proposed in Bluetooth). The protocol is degree-independent; that is, the number of channels is independent of the network degree. The protocol requires clock synchronization, which makes it difficult to apply in a large-scale network; however, this might not be a problem for PAN, as these are typical small sized networks.

7.2.6 Use of Multiple Antennas for MANET

7.2.6.1 Multiple Antennas Enhance Performance and/or Capacity

Antenna arrays, mainly used at the origin in radar applications, have been used for wireless communications for about 40 years now, starting with the use of directional antennas and spatial signal processing for direction of arrival (DOA) determination. Quite rapidly, in the late 1980s, the use of space-time processing [19] was proposed as a way to benefit from space and time diversity.

Indeed, both space and time diversity are present in the wireless channel, as the signals sent in the air bounce on all sorts of obstacles and arrive at the receiver from different directions and at different times. This diversity is best characterized by the *diversity gain*, which is equal to the number of independent replicas of the data signals that are combined, and transforms a Rayleigh channel (which has a BER proportional to the inverse of the SNR: BER := SNR^{-1}) in a channel that offers a performance enhancement given by BER := SNR^{-n}. Diversity (spatial and/or space-time diversity) was then the main motto for research, although the premises of multiple input multiple object (MIMO) systems could be seen in the literature about multiuser communications (a.o., spatial division multiple access). Finding simple and inexpensive ways to implement this diversity led to the first space-time code, known as the Alamouti code [20], which started a new research area. The idea of Alamouti was to put multiple antennas at the receiver rather than at the transmitter, as this could lead to an inexpensive way of providing diversity by putting multiple antennas in a base-station (which has been adopted in 3G).

Quite naturally, researchers started to put multiple antennas both at the transmitter and at the receiver. By doing this, they provided a higher order of diversity ($MT \times MR$, where MT and MR are the number of antennas at the transmitter and at the receiver, respectively), but more importantly discovered the possibility of using spatial multiplexing (SM) (or rather re-discovered, as signal separation [21] was already an active research area for more than 10 years).

Spatial multiplexing, whose best known avatar is V-BLAST (which stands for Vertical Bell Labs Layered Space-Time Architecture) [22], created many new possibilities: by multiplying the number of antennas by N at both the transmitter and receiver, it was possible to transmit N times more data without expanding the bandwidth nor the emitted power, and only at the expense of some coding and signal processing; N is then denoted as the multiplexing gain. This discovery led to a formidable research activity, both on the algorithmic and the information theory aspects, and it is now quite well understood, even if there is still a lot to do both from a theoretical and practical point of view. One of the fundamental questions was how much multiplexing and diversity gain is simultaneously achievable? This was answered by Zheng and Tse [23].

Both diversity and SM benefits can be easily implemented in cellular systems, as all communications are based on single links—that is, an uplink and a downlink. An example of this is the use of cell-diversity and enhanced soft handover. In mobile ad hoc networks, as they are not based on a centralized scheme, applying the above techniques is still a challenge, both from a networking and a hardware point of view.

One of the first answers to the use of antenna arrays in mobile ad hoc networks is the so-called directional MAC (DMAC) protocol proposed by Ko and Vaidya [24], in which they implement a modification to the 802.11 MAC to benefit from the directional antennas, which still solves the hidden and exposed terminal problems.

Another interesting approach is the multipacket reception approach by Tong et al. [25], in which the receiver is able to receive more than one packet at a time, possibly separating them, and hence, there is no real collision for these packets. A third approach is cooperative diversity [26], in which space and multiuser diversity is exploited by coordinated transmission from collaborating radios. Here, multiple copies of the sent signal, possibly coming from different transmitters (which are asked for it), are combined optimally at the MAC and/or the PHY level.

7.2.6.2   MAC Protocols for Multiple Antenna–Based Ad Hoc Networks

Although MAC protocols for multiple antenna–based ad hoc networks are a recent topic, there are many of these protocols now around. We focus on two typical examples: the first one is based on directional antennas, and the other benefits from the spatial multiplexing MIMO approach.

Directional antennas were recognized to be useful in ad hoc networks by Lau, Pronios, and others in the early 1990s, followed by Vaidya, Nasipuri, and Ramanathan in the late 1990s. Nasipuri and Vaidya introduced the principles of directional-MAC protocols and benchmarked them against information theoretic results.

7.2.6.3   Information Theoretic Results

The capacity of ad hoc networks has been addressed by Gupta and Kumar [27]. In this paper, they come to the conclusion that, under the best possible placement of nodes, and assuming a node that transmit W bits/sec, an ad hoc network could not provide a per-node throughput of more than $O(W/\sqrt{n})$—or $O(W/\sqrt{n \log n})$) for a random network—where n is the number of nodes in the network. Hence, the global capacity of the network is roughly $nO(W/\sqrt{n}) = O(W\sqrt{n})$.

The reason for this dramatic decrease in per-node capacity is due to the following:

- The silenced zone around a transmitter (determined by the RTS/CTS scheme) prevents nodes to transmit.
- To allow dense spatial reuse, this silenced zone is small; hence, communication is confined to nearest neighbors.
- If the silenced zone is minimized, the number of hops necessary for a communication in a large network may in general grow with $\sqrt{n}$.

From these facts, one can see that a packet is transmitted on the order of $\sqrt{n}$ times, and hence, the per-node throughput of $O(W/\sqrt{n})$ (for a rigorous proof, see [28]).

Mobility, relaying schemes, and hybrid (wireless ad hoc/wired backbone) networks can break this capacity, but directional antennas can also yield a substantial improvement in the capacity.

To make a long story short, consider the following (simplified) directional antenna, modeled as a sector characterized by the transmission/reception range r and the beamwidth α for transmission or β for reception. Consider a network with directional transmission and omnidirectional reception. Intuitively, we can argue that the transmitter will cause interference only in the portion of the initial circle that is covered by the sector (i.e., is $2\pi/\alpha$); hence, the directional transmission scales the capacity by $2\pi/\alpha$ (here we have assumed some form of power control—i.e., there is no antenna gain for the directional antenna—and we have assumed a random network). In the case of an arbitrary network, the scaling is found to be $\sqrt{(2\pi/\alpha)}$.

Consider now a network with directional transmission and reception (Figure 7.6). The amount of interference will be lower (still under the assumption that the ranges are the same in both omnidirectional and directional cases). Indeed, when node B receives data from node A, node B will be allowed to establish a communication with a node that is not in the direction of A. Like for the previous case, we have a gain of $2\pi/\alpha \cdot 2\pi/\beta = 4\pi^2/(\alpha\beta)$.

The conclusions globally hold for more complicated antenna models, even if the reasoning can not be asymptotically extended (i.e., there is no hope to get infinite capacity when using infinitely narrow beamwidths).

7.2.6.4 Problem Formulation

We place ourselves in a network with two types of antennas:

1. Omnidirectional antennas;

2. Directional antennas, modeled by their beamwidth and antenna gain G. Directional MACs have been developed with IEEE 802.11 as a base, trying to benefit from the following two characteristics:

   - *High spatial reuse:* IEEE 802.11 limits spatial reuse by silencing all nodes in the transmission range, and the use of multiple antennas enables the silence region to be limited, hence a higher spatial reuse.

   - *High transmission range:* Directional antennas have typically a gain G, which can be equal to $2\pi/\alpha$ (for the same total emitted power); hence, transmission range is higher and one can reach the destination in fewer hops.

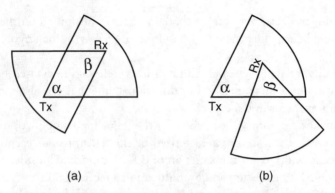

Figure 7.6 Directional antennas at both receiver and transmitter: (a) the receiver will be unable to transmit; and (b) the receiver will be allowed to transmit.

The basic DMAC [24] addresses the spatial reuse, and the multihop RTS MAC (MMAC) addresses the higher transmission range.

7.2.6.5 Basic DMAC

Basic DMAC relies on the knowledge of the transceiver profiles of the node and of its neighbors (i.e., characteristics of their antennas and mode of operation). The basic DMAC consists of two mechanisms similar to that of IEEE 802.11 (for transmitter node T and receiver node R):

1. *Channel reservation:* Channel reservation is based on an RTS/CTS mechanism where:
   - RTS is sent directionally; indeed, node T knows in which direction R is. Moreover, node T must check its directional NAV (DNAV) table to find out if it is allowed to transmit in R's direction. If it is allowed, he enters the back-off process and transmits his RTS at the end of this process.
   - The receiver listens omnidirectionally, and determines the DOA of the RTS. The other nodes update their respective DNAVs and enter the silenced region.
   - The receiver answers with a CTS directionally.
   - The data exchange is done directionally.
   - The acknowledgment is done directionally.
2. *Directional NAV table:* To benefit from the directional antennas, the NAV table also has to be directional. Indeed, to allow a node to transmit in a certain direction, it must maintain a table with the nonallowed directions. Consider the scenario depicted in Figure 7.7.

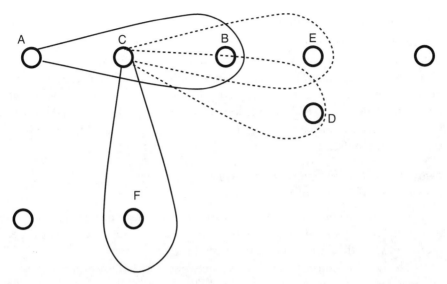

Figure 7.7 A transmits to B; hence, C may not transmit to D nor E, but may transmit to F (dotted line means not allowed).

In this scenario, A transmits to B; C overhears the RTS and CTS and updates its DNAV table. Hence, C will defer its communication to node E. Note also that even if D is not in the direction of B, C may not send to D; hence, the BDMAC protocol must take the beamwidth of the directional antennas into account. On the other hand, the direction of F is free (that is, the DNAV table does not disallow the direction of F) and the communication between C and F can take place.

The main drawback of the basic DMAC is that it introduces new hidden terminal and deafness problems.

7.2.6.6 Multihop RTS MAC

One of the possible answers to the drawbacks of the basic DMAC is the multihop RTS MAC protocol, which basically decouples the RTS propagation from the CTS and actual data propagation, transmitting the RTS in a multihop fashion.

Choudury et al. performed simulations with directional antennas featuring a 45° beamwidth and a gain of 10 dB. For various topologies involving 25 nodes, the aggregate throughput gain (when using the MMAC) is about 4 to 5 with various configurations; the basic DMAC suffers too much from its drawback to be efficient.

Note that a gain of 4 to 5 is not so far from the gain of 8 that was predicted by information theoretic arguments. Note also that, the gains are similar to the gain obtained with multipacket reception, as shown in the next section.

7.2.6.7   Multipacket Reception

A natural way to benefit from the advances of MIMO signal processing is the so-called multipacket reception (MPR) introduced by Zhao and Tong [23, 25]. Using multiple antennas, the receiver can possibly get multiple packets simultaneously. The question is: how many concurrent packets should the MAC ask for? Tong denotes an MPR node as a wireless node capable of correctly receiving signals from multiple transmitters. This can be performed by any form of signal separation (see a recent overview of this in [28]).

Networks with MPR Nodes and the Multiqueue Room Service (MQRS) Protocol

The goal of this algorithm is twofold: first, it tries to maximize the network capacity (defined as the maximum number of packets to send per slot and per user), and second, it tries to guarantee a certain QoS.

The network considered is a slotted network, with a finite number of users that may have different QoS requirements. The QoS of a given user is only characterized by its average packet delay at the heaviest traffic load.

The user transmits equal-sized packets in one time slot and generates a packet with probability p. Users are partitioned into L groups, according to their QoS, and each group contains M^l users (for a total of $M = \Sigma\, M^l$), with a maximum delay requirement for the heaviest load (i.e., for $p = 1$) of d^l.

The first major result is that Zhao and Tong proved the existence of a MAC protocol that can serve heterogeneous QoS requirements (but in the case of a centralized protocol). Moreover, as their proof is constructive, they propose the MQRS protocol that implements this capacity maximization and guarantees the QoS. This protocol is a based on a central controller that manages L parallel queues:

- One queue per QoS class;
- Two rooms per queue: one room is an access set (i.e., the set of users that want access to the channel in the current slot) and one is the waiting room (hosting unsuccessful users);
- Two parameters: M_i, the number of users in the access set, and q, the probability to choose a user from queue 1.

Equipped with these parameters, queues, and rooms, we can explain the basic structure of the MQSR protocol in the case of L = 2 classes of QoS (for the quite complex determination of the parameters, see [23]). Following Figure 7.8, the MQSR protocol can be divided in four steps (A through D):

    A. Users are put in the queue, in a given order.

    B1. Controller determines the number of users in the access set K, as well as q.

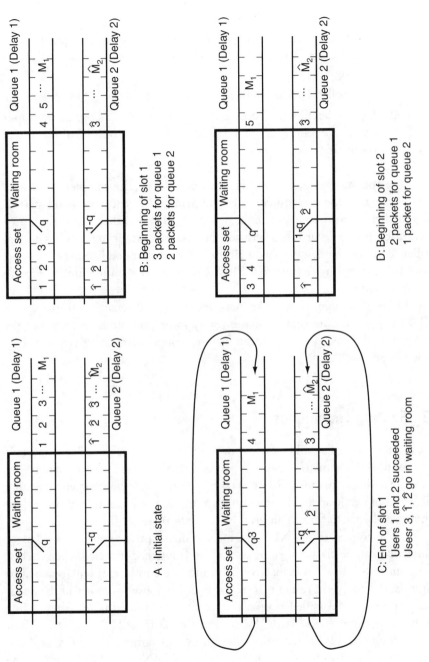

Figure 7.8 Procedure for MQSR protocol: the case of two different delay requirements.

B2. Controller chooses the users that will be in the access set (takes a user from queue 1 with probability q and from queue 2 with probability 1 – q).

C. Controller determines which users have been successful, putting successful users at the end of the queue and unsuccessful users in the waiting room.

D1. Controller determines the number of users in the access set K.

D2. Controller fills the access set like in step B2, taking as the input queue the waiting room followed by the queue outside of the service room, and goes to step C.

The apparent simplicity of this protocol hides a quite complex controller, in the sense that the determination of the parameters requires a good statistical knowledge (estimation) of the user's state. On the other hand, the introduction of the waiting room ensures fairness at quite a low cost. The real challenge here is the design of a distributed algorithm for ad hoc network operation. It is interesting to note that MQSR is shown (by simulation) to achieve the capacity.

Other like algorithms, like network-assisted diversity multiple access (NDMA) by Tsatsanis, offer similar capacity gains, and can be extended in the framework of MIMO communications and cooperative diversity, as will be shown in Section 7.5.

7.3 Routing Techniques

The primary goal of an ad hoc network routing protocol is the correct and efficient route establishment between a pair of nodes so that messages may be delivered in a timely manner. Route determination should be done either with a minimum of overhead and bandwidth consumption or in a minimal time; algorithms usually try to find a compromise between these conflicting wishes.

In the context of MANET, the focus is on adaptive and distributed algorithms that can be classified either as reactive (route is discovered when needed to minimize overhead) or proactive (all routes are continuously maintained to minimize delay). In the rest of this section, we give a short overview of the major routing protocols (classified as proactive, reactive, and hybrid), focusing mainly on the unicast protocols. As the focus of this book is on WPANs, we will not give thorough details of all routing algorithms, but rather an intuitive explanation of major protocols that could be used in WPANs (like DSR, used in the PACWOMAN project).

7.3.1 Proactive Routing Protocols

Proactive unicast protocols attempt to maintain routing tables in all nodes, with consistent and fresh routing information to all destinations.

7.3.1.1 Destination Sequenced Distance Vector

Destination Sequenced Distance Vector (DSDV) [29] is an adaptation of the conventional distance vector routing protocol (the distributed version of the famous Belman-Ford algorithm) to ad hoc networks. The major enhancement is the avoidance of loops. In DSDV, packets are routed between nodes of an ad hoc network using routing tables stored at each node. Each routing table lists, for each destination,

- The next hop;
- A cost metric (e.g., the number of hops);
- A destination sequence number, created by the destination and used to avoid loops.

To maintain the consistency of the routing tables, DSDV uses both periodic and triggered routing updates (triggered update is used upon topology change, to propagate the new topology as fast as possible). Upon routing update, each node increments and appends its sequence number, so that upon reception of a routing update, the route entry is tagged with the new sequence number. DSDV claims the following properties:

- Loop-free at all instants;
- Dynamic, multihop, self-starting;
- Low memory requirements;
- Quick convergence via triggered updates;
- Routes available for all destinations;
- Fast processing time;
- Reasonable network load;
- Minimal route trashing;
- Intended for operation with up to 100 mobile nodes, depending on *mobility factor*.

The major advantage of DSDV is that it provides loop-free routes at all instants maintaining a moderate memory requirement. The major drawback, besides the overhead due to periodic updates, is that the optimal values for the parameters (periodicity of updates) are difficult to determine, leading to possible route fluctuations and spurious advertisements and waste of bandwidth.

7.3.1.2   Wireless Routing Protocol

The Wireless Routing Protocol (WRP) [30] is a path-finding algorithm that is also based on a distance vector algorithm. To avoid the looping problem present in the (original) distance vector algorithm, WRP includes the length and the second-to-last hop (predecessor) information of the shortest path to each destination.

Due to changes in radio connectivity or jamming, the messages may be lost or corrupted. After receiving an update message free of errors, a node is required to send a positive ACK indicating that it has good radio connectivity and has processed the update message. A node can send a single update message to inform all its neighbors about changes in its routing table because of the broadcast nature of the radio channel; however, each neighbor sends an ACK to the originator node.

A periodic update message is sent without any changes of routing table if no recent transmissions of routing table updates or ACKs are received, to ensure the connectivity with the neighbor. If a node fails to receive any type of message from a neighbor in a specified amount of time, the node must assume that connectivity with that neighbor has been lost.

The major advantage of WRP is that it reduces temporary looping by using the predecessor information to identify the route.

In WRP, each node is required to maintain four routing tables. This can lead to substantial memory requirements, especially when the number of nodes in the network is large. Furthermore, if there are no recent packet transmissions from a given node, the WRP requires the use of hello packets. The hello packets consume bandwidth and disallow a node to enter sleep mode.

7.3.1.3   Hierarchical Routing

The main disadvantage of the WRP and DSDV algorithm families is that that all nodes maintain routing tables for all destinations. As such, these algorithms do not scale to large networks, due to the excessive overhead and routing table memory.

To alleviate this problem, the use of hierarchical routing algorithms in [31], namely fisheye state routing (FSR) and hierarchical state routing (HSR), allow a decrease of memory requirements and routing table updates. The first algorithm assumes that the information of the nearest nodes is known with much more detail than that of far-away nodes; this enables the minimization of routing tables and also of the bandwidth consumed by the routing table exchanges. The second algorithm is based on a clustered hierarchy of the network, also lowering the routing table sizes, but also the number of updates needed.

7.3.2 Reactive Protocols

Reactive protocols try to minimize the resources needed, especially the overhead due to routing table exchanges, by determining a route only if needed. Hence, when a node wants to send a packet to a given destination, it starts a routing discovery algorithm and maintains this route (including router repair) as long as it is needed for data communication.

7.3.2.1 Dynamic Source Routing Protocol

Dynamic source routing (DSR) [32] allows a node to dynamically discover a route across multiple network hops to any destination. Source routing means that each packet in its header carries the complete ordered list of nodes through which the packet must pass. DSR uses no periodic routing messages, thereby reducing network bandwidth overhead (at least in relatively low to moderate traffic situations), conserving battery power and avoiding large routing updates throughout the ad hoc network. Instead, DSR relies on support from the MAC layer (the MAC layer should inform the routing protocol about the link failure) and is based on the flooding of control packets for route discovery and route maintenance.

If a source node wants to communicate with a destination node, it checks in its cache if it has a nonexpired route to the destination (a route expires if the node receives a route error packet saying that one of the links in the route has gone down). If it does, it sends the packet with this route, hop-by-hop, contained in the header field of the packet. If the source does not have a route to the destination, route discovery is initiated. This involves broadcasting a route request (RREQ) packet to its neighbors, adding the source and destination addresses as well as a unique identifier to the packet. The nodes that receive this packet will check if they have a nonexpired route to the destination, or if they are the destination. If this is not the case, they will append their own address to the packet header field and forward the packet on their outgoing links. In order to limit the message complexity in the system, a node will only forward a packet if it has not already seen the packet before (i.e., it does not appear in the header field).

If the node is the destination, or it has a nonexpired route to the destination, it will append its own address to the header and return the packet to the source [as a route reply (RREP) packet], either by backtracking through the nodes indicated by the header field, or by initiating its own route discovery back to the source node.

Route maintenance is achieved through route error packets and acknowledgments. When a link goes down, a route error packet (RERR) is generated. When a route error packet is received, the hop in error is removed from the node's routing cache, and all routes are truncated at that point.

DSR makes very aggressive use of source routing and route caching. Several additional optimizations have been proposed and have been evaluated to be very effective by the authors of the protocol, as follows:

- *Salvaging:* An intermediate node can use an alternate route from its own cache when a data packet meets a failed link on its source route.
- *Gratuitous route repair:* A source node receiving an RERR packet piggybacks the RERR in the following RREQ. This helps clean up the caches of other nodes in the network that may have the failed link in one of the cached source routes.
- *Promiscuous listening:* When a node overhears a packet not addressed to itself, it checks whether the packet could be routed via itself to gain a shorter route. If so, the node sends a gratuitous RREP to the source of the route with this new, better route. Promiscuous listening also helps a node to learn different routes without directly participating in the routing process.

The main advantage of DSR is its use of source routing. This means that intermediate nodes do not have to maintain routing information, thus lowering the bandwidth waste as well as saving battery power. DSR also avoids the need for periodic routing messages. This advantage is very important when mobility is low, as no routing maintenance is needed, and for low traffic situations, by enabling nodes to stay in sleep mode.

Moreover, as the routes are carried with the data packets, the intermediate nodes can learn the routes by inspecting these data packets, which also contribute to the lowering of the overhead. On the other hand, these mechanisms allow (sometimes) faster reaction to topology changes than link-state/distance-vector based protocols.

It is also worth noting that bidirectionality of the links are not necessary (as the RREPs could take different routes than the RREQs)

As for WRP and DSDV, DSR does not scale well to large networks, as the control packets and the message packets become larger as the path grows. Hence, some hierarchical routing algorithms are also needed here [31].

7.3.2.2 Ad Hoc On-Demand Distance Vector Protocol

One of the main disadvantages of DSR is its potentially large header that contains the source routes. The Ad Hoc On-Demand Distance Vector (AODV) protocol attempts to solve this by using routing tables at the nodes, so that the headers are kept to a minimum.

In AODV [33], the source node broadcasts a RREQ packet. This RREQ packet contains a hop count, the destination and source's IP address, as well as a sequence number for the source and a sequence number for the destination.

A route to the destination node is determined either by the RREQ packet arriving at the destination node itself or at some intermediate node that has a valid route to the destination. A valid route is a route whose entry for the destination sequence number is at least as great as that contained in the RREQ to ensure freshness of the route information. The route is made available to the source node by sending a RREP packet back to the source node. In order to achieve this backtracking of the RREP packet, each node maintains a cache of a route back to the originator of each RREQ.

Nodes monitor link status to the next hop in active routes. When a link goes down, the node sends a RERR packet to other nodes indicating the destinations that are now unreachable due to the link failure. In order to achieve this, each node needs to maintain a *precursor list* that indicates what nodes are likely to have been affected by the link failure. This precursor list is most easily obtained when dealing with the backtracking associated with the routing of an RREP packet. The use of this back-link also means that AODV, unlike DSR, assumes bidirectional links. Sequence numbers are used together with hop-counts to avoid loop formation.

AODV needs to keep track of the following information for each route table entry:

- *Destination IP address:* IP address for the destination node;
- *Destination sequence number:* sequence number for this destination;
- *Hop count:* number of hops to the destination;
- *Next hop:* the neighbor that has been designated to forward packets to the destination for this route entry;
- *Lifetime:* the time for which the route is considered valid;
- *Active neighbor list:* neighbor nodes that are actively using this route entry;
- *Request buffer:* makes sure that a request is only processed once;

AODV, like DSR, is responsive to changes in topology and ensures that no loops are formed. AODV improves on DSR by using bandwidth more efficiently (by minimizing the network load for control and data traffic) and by improving scalability. A definite advantage is also the possible extension of AODV to support both unicast and multicast routing.

7.3.2.3 Temporally Ordered Routing Algorithm

DSR and AODV are based on the flooding of control packets, yielding a large overhead for high-traffic/high-mobility networks. Link reversal algorithms avoid this flooding. The aim of link reversal algorithm [here we will present the main advantages of the temporally ordered routing algorithm (TORA) [34]] is to build a directed acyclic graph (DAG) for each destination, with the destination being the only sink of the graph.

The DAG is constructed first by flooding control packets from the source to the whole network. In a second phase, (logically) directed links are established between adjacent nodes, and the link reversal algorithm (which we will not detail) results in the formation of the DAG. Hence, the construction of the initial DAG generates a lot of control packet traffic. DAG maintenance takes advantage of the link reversal method. Indeed, if a link is broken, the link reversal method ensures that the control packet traffic is (mostly) limited to the neighborhood of the broken link. Moreover, the structure of a DAG is such that more than one route can be found to a given destination.

Classical link reversal algorithm can lead to deadlocks (the algorithms continue to search a DAG indefinitely) in the case of partitioned networks (the network is partitioned in two subnetworks without any link between them). TORA uses a modified link reversal method (by detecting unproductive link reversals) to detect partitions and solve the infinite routing determination loop problem.

One of the disadvantages of TORA (and all DAG-based protocols) is that routing is not optimized (paths may not be the shortest).

7.3.3 A Hybrid Protocol: Zone Routing Protocol

The Zone Routing Protocol (ZRP) (see [35] and references therein) is a hybrid of a reactive and a proactive routing protocol. It divides the network into several routing zones. A zone is a local region defined by a single parameter called the zone radius, which is measured in hops. Nodes proactively maintain routing information for nodes within their zones and reactively discover routes for nodes outside their zones. As such, ZRP can also be considered as a sort of hierarchical routing protocol. The two routing mechanisms are referred to as the Intrazone Routing Protocol (IARP) and Interzone Routing Protocol (IERP), respectively.

The protocol used inside the routing zone is not defined and can include any number of proactive protocols, such as distance vector or link-state routing. Different zones may operate with different intrazone protocols as long as the protocols are restricted to those zones.

When a node needs to send packets to a destination, it first checks to see if the destination is in the same zone. If so, the path to the destination is known (through IARP) and the node delivers them accordingly. If the destination is not within the source's routing zone, the source broadcasts (or bordercasts) a query to all of its peripheral nodes (which are those whose minimum distance from the node in question is the zone radius), which in turn forward the request if the destination node is not found within their routing zone.

This procedure is repeated until the requested node is found and a route reply is sent back to the source indicating the route. IERP uses Bordercast Resolution Protocol (BRP), which is included in ZRP.

Since this is a hybrid between proactive and reactive schemes, this protocol use advantages from both. Routes can be found very quickly within the routing zone, while routes outside the zone can be found by efficiently querying nodes in the network.

7.4 Power Management

7.4.1 Power-Saving MAC Protocols

The most well-know power-saving MAC protocols are simply the power control implemented in cellular networks, where the power is adjusted so as to have a predetermined received power at the base station in the uplink (and at the mobile in the downlink), based on received power measurements.

In ad hoc networks, this simple protocol can be implemented in an 802.11-like protocol by sending the RTS/CTS packets at a maximum power and using the received power indication to adjust the data packets at the minimum power needed for proper reception. Note that adoption of this mechanism would lead to the network capacity derived in [27].

To enable smart adoption of this scheme, and thus allow the achievement of network capacity, the goal of the power control is to reduce the node degree while maintaining a connected network. Different power levels can be allocated to different destinations, and the knowledge of the topology and position of the nodes becomes a major parameter for power-aware MAC algorithms. References [36–39], for example, rely on various forms of position and topology knowledge to adjust power and/or control interference. In practical cases, a power saving of a factor two has been achieved, and this is a good incentive to incorporate these approaches in future WPAN/WLAN standards.

7.4.2 Power-Aware Routing

The first authors to consider power-aware routing were Singh et al. [40]. They introduced new metrics for path selection, including energy consumed per packet, network connectivity duration, node power variance, as well as maximum node cost (taking into account the fixed energy cost to maintain the node).

Proactive algorithms [e.g., Power Aware Multiple Access Protocol (POMAS)] were the first to use these new metrics, using the power cost instead of delay or hop count as link weight. Taking into account that these proactive networks imply the periodic exchange of routing information, it has been shown that for networks with low mobility these schemes consume more power than reactive protocols. Hence, reactive power optimized algorithms like power-aware routing optimization (PARO) [41] were proposed, using power as a link metric while lowering the exchange of control packets.

Naïve use of these metrics can lead to unequal node power exhaust, which in turn leads to some node breakdown and to network partition, which is not wanted. To alleviate this problem, one may combine the power metrics with the use of directional antennas and Dijkstra-like algorithms to minimize the energy consumed per packet [36].

From a performance perspective, the use of power-aware routing combined with simple power-aware MAC routing protocols achieve significant power reduction, but often at the expense of a loss in throughput, except for specific assumptions about the context (low mobility and symmetric channels, for which special algorithms can be designed). Moreover, in the case of highly populated networks, where interference is important (and where power control is absolutely necessary), these protocols have not yet been thoroughly evaluated and so a lot of open issues need to be investigated.

7.5 Cooperation Diversity: The New Revolution

Cooperation diversity exploits the spatial diversity of the relay channel by allowing different mobile terminals to cooperate. In this context, the formation of *virtual antenna arrays* is often envisaged.

Cooperation diversity was introduced by Laneman, Tse, and Wornell in [26] and is illustrated by the following example. Consider four terminals (T_1 to T_4), in which T_1 and T_2 respectively transmit to T_3 and T_4. In classical communications, T_1 and T_2 would send their data one after the other (in a TDMA fashion). Cooperation between the two senders could be used by distributed space-time coding, considering each sender as part of a virtual antenna array. The main difficulty of this approach is the need for synchronization as well as the sharing of the information at both transmitters. Following [26], we focus here on *orthogonal cooperative diversity* (which is indeed the most relevant, as wireless systems are unable to simultaneously receive and transmit at the same frequency/code), and introduce *amplify-and-forward* and *decode-and-forward* relaying strategies. In the amplify-and-forward strategy, the relay retransmits a version of its received signal by merely amplifying it. In the decode-and-forward strategy, the relay decodes the message and re-encodes it.

The basic relaying scheme is illustrated in Figure 7.9

The scenario in Figure 7.9 resembles a simple two-hop relaying, but the revolution lies in the fact that, at the destination, the two signals impinging have gone through different channels. Hence, at the destination, they can be

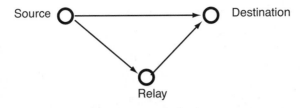

Figure 7.9 Simple cooperative diversity.

combined in a proper way, and the destination can take advantage of the spatial diversity (signals coming from different directions) and the time diversity (signals coming from the source and relay arrive at different moments, in a TDMA way, for example).

Cooperation diversity combines multihop networking and space-time diversity (the source/relay can be considered as a virtual antenna array with two elements). The latter offers great benefits in terms of performance and power savings.

Combining cooperative diversity with multiple antennas at both ends of the link in ad hoc networks promises to offer both great capacity (using multiuser diversity) and performance (using space-time diversity) enhancements—at the cost of a large system complexity. Although little work has been done on this topic, Madueño and Vidal have explored this approach in [42], combining NDMA and cooperative diversity. They show large capacity gains for a large set of scenarios and, in particular, show that for nonhomogeneous networks, the capacity tends to be fairly distributed among nodes.

7.6 Conclusions

MANET has evolved into a mature research topic and is a major part of WPAN's deployment. Even if routing algorithms and power awareness have been thoroughly studied, the advent of new physical layer–oriented evolutions promise dramatic capacity improvements that have yet to be taken over in the higher layers. Moreover, the use of cooperation diversity profoundly modifies the MANET paradigm, making network assisted diversity (the networking counterpart of multiuser diversity) a new challenge that is yet to be assessed in both theory and in practice.

References

[1]   Corson, M. S., and J. Macker, Mobile Ad hoc Networking (MANET), "Routing Protocol Performance Issues and Evaluation Considerations," Request For Comments 2501, Internet Engineering Task Force, January 1999.

[2]   http://www.ietf.org/html.charters/manet-charter.html.

[3]   McQuillan, J. M., and D. C. Walden, "The ARPA Network Design Decisions," *Computer Networks*, Vol. 1, pp. 243–289.

[4]   Kahn, R. E., et al., "Advances in Packet Radio Technology," *Proceedings of the IEEE*, Vol. 66, No 11, November 1978.

[5]   Gupta, P., and P. R. Kumar, "The Capacity of Wireless Networks," *IEEE Trans. on Info. Theory*, Vol. 46, No. 2, March 2000, pp. 388–404.

[6]   Gomez, J., et al., "PARO: Conserving Transmission Power in Wireless ad hoc Networks," *IEEE 9th International Conference on Network Protocols (ICNP'01)*, Riverside, California. November 2001.

[7]   Toh, C. -K., "Maximum Battery Life Routing to Support Ubiquitous Mobile Computing in Wireless Ad Hoc Networks," *IEEE Communications*, June 2001.

[8]   Markopoulos, A., et al., "PACWOMAN Power Aware Communications for Wireless OptiMised Personal Area Network Contract No IST-2001-34157 D 6.1—Requirement Analysis for Security Mechanisms," July 2003.

[9]   Gavrilovska, L., et al., "Future Adaptive Communication Environment D 1.1—Status Report on State-of-the-Art in Short-Range Networking," September 2002.

[10]   Toh, C.-K., *Ad Hoc Mobile Wireless Networks: Protocols and Systems*, Englewood Cliffs, NJ: Prentice Hall, 2002.

[11]   Karn, P., "MACA a New Channel Access Method for Packet Radio," *Proceedings of the 9th ARRL/CRRL Amateur Radio Computer Networking Conference*, September 1992.

[12]   Bharghavan, V., et al., "MACAW: A Media Access Protocol for Wireless LAN's," *Proc. ACM SIGCOMM*, London, U.K., August 1994, Vol. 1, pp. 212–225.

[13]   Fullmer, C. L., and J. J. Garcia-Luna-Aceves, "Solutions to Hidden Terminal Problems in Wireless Networks," *Proceedings ACM SIGCOMM*, Cannes, France, September 1997.

[14]   Muir, A., and J. J. Garcia-Luna-Aceves, "Group Allocation Multiple Access with Collision Detection," *INFOCOM '97, Sixteenth Annual Joint Conference of the IEEE Computer and Communications Societies, Proceedings IEEE*, April 1997.

[15]   Tobagi, F. A., and L. Kleinrock, "Packet Switching in Radio Channels: Part II—The Hidden Terminal Problem in Carrier Sense Multiple Access Modes and the Busy-Tone Solution," *IEEE Transactions on Communications*, Vol. 23, No 12, 1975.

[16]   Lin, C. R., and J.-S. Liu, "QoS Routing in Ad Hoc Wireless Networks," *IEEE Journal on Selected Areas in Communications*, Vol. 17, No. 8, August 1999, pp. 1426–1438.

[17]   Tseng, Y.-C., et al., "A Multi-Channel MAC Protocol with Power Control for Multihop Mobile Ad Hoc Networks," *IEEE*, 2001, pp. 419–424.

[18]   Nasipuri, A., J. Zhuang, and S. R. Das, "A Multichannel CSMA MAC Protocol for Multihop Wireless Networks," *Proceedings of WCNC'99*, September 1999.

[19]   Paulraj, A., and C. B. Papadias, "Array Processing for Mobile Communications," in *Handbook on Signal Processing*, CRC Press, 1997.

[20] Alamouti, S., "Space Block Coding: A Simple Transmitter Diversity Technique for Wireless Communications," *IEEE Journal on Selected Areas in Communications*, Vol. 16, October 1998, pp. 1451–1458.

[21] Comon, P., "Independent Component Analysis, a New Concept?" *Signal Processing*, Vol. 36, No. 3, 1994, pp. 287–314.

[22] Foschini, G. J., "Layered Space-Time Architecture for Wireless Communicationsin a Fading Environment when Using Multi-Element Antennas," *Bell Labs Tech. J.*, Vol. 1, No.,2, 1996, pp. 41–59.

[23] Zhao, Q., and L. Tong, "A Multiqueue Service Room MAC Protocol for Wireless Networks with Multipacket Reception," *IEEE/ACM Trans. on Networking*, Vol. 11, No. 1, February 2003, pp. 125–137.

[24] Ko, Y.-B., B. Shankarkumar, and N. H. Vaidya, "Medium Access Control Protocols Using Directional Antennas in Ad Hoc Networks," *INFOCOM, Nineteenth Annual Joint Conference of the IEEE Computer and Communications*, 2000, pp. 13–21.

[25] Tong, L., Q. Zhao, and G. Mergen, "Multipacket Reception in Random Access Wireless Networks: From Signal Processing to Optimal Medium Access Control," *IEEE Communications Magazine*, Vol. 39, No. 11, November 2201, pp. 108–112.

[26] Laneman, N., D. Tse, and G.Wornell, "Cooperative Diversity in Wireless Networks: Efficient Protocols and Outage Behavior," *IEEE Trans. on Inf. Theory*, Vol. 50, No. 12, December 2004, pp. 3062–3080.

[27] Gupta, P., and P.R. Kumar, "The Capacity of Wireless Networks," *IEEE Trans. on Info. Theory*, Vol. 46, No. 2, March 2000, pp. 388–404.

[28] Giannakis, G., et al., *Signal Processing Advances in Wireless Communications*, Englewood Cliffs, NJ: Prentice-Hall, 2001.

[29] Perkins, C. E, and P. Bhagwat, "Highly Dynamic Destination-Sequenced Distance-Vector Routing (DSDV) for Mobile Computers," *Proceedings of SIGCOMM '94 Conference on Communications, Architectures, Protocols and Aplications*, August 1994, pp. 234–244.

[30] Murthy, S., and J. J. Garcia-Luna-Aceves, "A Routing Protocol for Packet Radio Networks," *Proceedings, First International Conference on Mobile Computing and Networking (ACM Mobicom)*, Berkeley, California, November 13–15, 1995.

[31] Iwata, A, et al., "Scalable Routing Strategies for Ad Hoc Wireless Networks," *IEEE Journal on Sel. Areas in Comm.*, Vol. 17, No. 8, August 1999, pp. 1369–1379.

[32] Johnson, D. B., and D.A. Maltz, "Dynamic Source Routing in Ad Hoc Wireless Networks," in *Mobile Computing*, T. Imielinski and H. Korth (Eds.), Kulwer, 1996, pp. 152–181.

[33] Perkins, C. E., and E. M. Royer, "Ad hoc On-Demand Distance Vector Routing," *Second IEEE Workshop on Mobile Computing Systems and Applications, 1999, Proceedings, WMCSA '99*, February 25–26, 1999, pp. 90–100.

[34] Park, V. D., and M.S. Corson, "A Highly Adaptive Distributed Routing Algorithm for Mobile Wireless Networks," *Proceedings of IEEE Conference on Computer Communications (INFOCOM)*, Kobe, Japan, April 1997, pp. 1405–1413.

[35]   Pearlman, M. R., and Z. J. Haas, "Determining the Optimal Configuration for the Zone Routing Protocol," *IEEE Journal on Selected Areas in Communications*, Vol. 17, No. 8, August 1999, pp. 1395–1414.

[36]   Spyropoulos, A., and C. S. Ragavendra, "Energy Efficient Communications in Ad Hoc Networks Using Directional Antennas," *Prox. IEEE INFOCOM Conf.*, April 2003.

[37]   Rodoplu, V., and T. Meng, "Minimum Energy Mobile Wireless Networks," *IEEE Journal on Spec. Areas in Comm.*, Vol. 17, No. 8, August 1999, pp. 1333–1344.

[38]   Wattenhofer, R., et al., "Distributed Topology Control for Power Efficient Operation in Multihop Wirleless Ad Hoc Network," *Proc. IEEE INFOCOM Conf.*, 2001, pp. 1388–1397.

[39]   Monks, J., V. Bhargavan, and W. M. Hwu, "A Power Controlled Multiple Access Protocol for Wireless Packet Networks," *Proc. IEEE Infocom Conf.*, 2001, pp. 219–228.

[40]   Singh, S., M. Woo, and C.S. Raghavendra, "Power Aware Routing in Mobile Ad Hoc Networks," *Proc. ACM MobiCom Conf.*, 1998, pp. 181–90.

[41]   Gomez, J., et al., "PARO: Supporting Dynamic Power Controlled Routing in Wireless Ad Hoc Networks," *ACM/Kluwer J. Wireless Networks*, Vol. 9, No. 5, 2003, pp. 443–60.

[42]   Madueño, M., and J. Vidal, "Joint Physical-MAC Layer Design of the Broadcast Protocol in Ad Hoc Networks," *IEEE Journal on Selected Areas in Communications+*, Vol. 23, No. 1, January 2005, pp. 65–76.

Selected Bibliography

Kurnz, M., A. Mugattash, and S.-J. Lee, "Transmission Power Control in Wireless Ad Hoc Networks: Challenges, Solutions, and Open Issues," *IEEE Network+*, September/October 2004, pp. 8–15.

Sendonaris, A., E. Erkip, and B. Aazhang, "User Cooperation Diversity—Part I: System Description," *IEEE Transactions on Communications+*, Vol. 51, November 2003, pp. 1927–1938.

Tang, Z., and J. J. Garcia-Lunes-Aceves, "Hop-Reservation Multiple Access (HRMA) for Ad Hoc Networks," *Proceedings of Infocom'99+*, October 1999, pp, 194–201.

Zhang, R., and M. K. Tsatsanis, "Network-Assisted Diversity Multiple Access in Dispersive Channels," *IEEE Trans. on Comm.+*, Vol. 50, No. 4, April 2002 pp. 623–632.

8

Security for Wireless Networks

8.1 Introduction

Until recently, the ability to communicate to anyone, anywhere, anytime was only a vision. By 2010 it is expected to be a reality [1, 2]. The rapid evolution of mobile and wireless communication has been fueled by the growing demand for mobile information services. Along the way, our expectations have also grown. Services unavailable a decade ago are taken for granted today. Voice and low rate data are only the beginning. Not only do we expect to be able to use our handsets of the future anywhere, we expect them to support an increasingly wide range of bandwidth-demanding information services and operating conditions. Handsets must be increasingly capable, flexible, and adaptable using software radio architecture.

With the growth of wireless and the Internet, an obvious merge of the technologies is taking place. Several standardizing committees, very often with representatives from same companies, have formed many standards. These standards can be divided into the following categories: satellite communications, cellular technology, WLAN, WPAN, cordless technologies, and fixed wireless access (FWA) [3–8].

The overall market demand is for connectivity, mobility, security, and performance. Wireless services can provide connectivity, mobility, and performance, but not good security, which has been the weakest link for wireless communications. In the past, standardization bodies devoted very little time and effort in developing very good security architecture [9–11] for any of the existing systems. Thus, together with the growth of Internet, there has been tremendous growth in the field of wireless communications. This is due to the obvious benefits of wireless, namely, increased mobility and decreased wiring complexity and also increased flexibility and ease in installation. This ease came at a cost to security. Wireless WAN (WWAN) technologies based on second generation (2G/2.5G) standards like GSM claimed to be the most secure network, but recently Israeli scientists cracked GSM mobile call security by using a special device that steals calls and impersonate callers in the middle of a call [12]. In Germany (on Thursday, October 2, 2003), hackers broke into GPRS billing. The scam is called the over-billing attack. It works quite simply because of a link from the Internet world (which is unregulated) to the normally tightly regulated GSM planet. "Network administrators face an exponential onslaught of attacks that to date have traditionally been confined to the world of wire line data," was the summary from *Check Point* [13, 14].

At present, security solutions are provided by the UMTS Subscriber Identity Module (USIM) algorithms for UMTS [15], IEEE 802.11i for WLAN security [16], and Bluetooth [17] security recommendations tailored for securing the traffic exchanged between user devices and access points. User requirements are not taken into account during initial stages of designing the services and applications, but by the combined influence from terminals and networks developed according to the current technology.

With the widespread use of smart mobile phones, PDAs, laptops, and the availability of cheap wireless networking hardware, there has been a mushrooming in the number of wireless access networks. Currently, such networks are operated by individual organizations (stakeholder or administrative domains) and are usually closed to users who belong to other network operators or stakeholders.

This chapter describes the wireless networks and its security issues. Security solutions and drawbacks of the existing standards are discussed in Section 8.2. Applications and services are analyzed in Section 8.3, and security requirements are reviewed in Section 8.4. Section 8.5 provides research challenges.

8.2 Heterogeneous Networks

Security in heterogeneous networks is one of the biggest technology challenges of the 21st century. However, like for many issues, everybody talk about it, but the ones who should feel the most affected do not seem to have detected the scale of the potential disaster. Figure 8.1 shows a heterogenous network for call setup, authentication, authorization, and payments applicable for various IP communications. A very heterogeneous network, contains different systems that are more or less widespread. The more the access networks, the worse the complexity, since not all systems are equal in front of adversity. This section discusses the security architecture and flaws of existing access technologies.

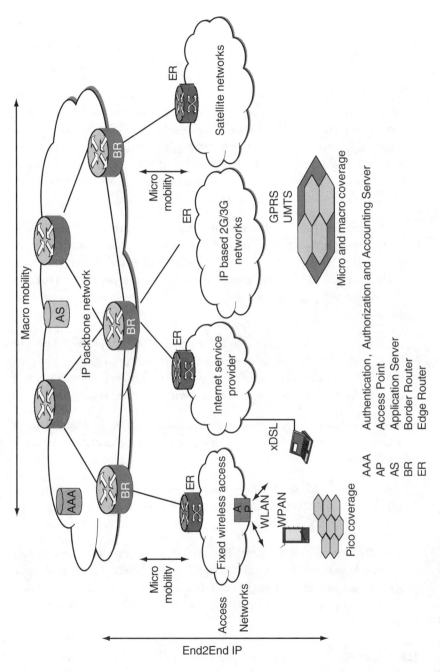

Figure 8.1 Heterogeneous networks.

8.2.1 Satellite Networks

Satellite networks [18, 19] are especially sensitive to security attacks because the transmitted data can be easily intercepted and corrupted. Thus, it is very important to provide the following strong security mechanisms:

- *Confidentiality:* This service protects data from passive attacks. It protects against unauthorized release of message content. It may also provide a protection against traffic flow analysis (e.g., source, destination, frequency, length). This service has to be provided (unless the user does not need any communication privacy—for example, free information broadcasting). This protection is needed because it is very easy for an attacker to intercept satellite communications.

- *Authentication:* This service guarantees that the communication is authentic. It assures the recipient of a message that the message is from the source that it claims to be from. It is necessary to have a strong authentication mechanism since it is very easy to impersonate a user on a wireless communication and because IP address fields cannot be trusted.

- *Integrity:* This service assures that the message (or the protected part of it) is received as sent. Packet integrity has to be cryptographically protected. An attacker can easily modify packets and create new packets if there is no integrity protection. This service has to be provided.

- *Nonrepudiation:* This service prevents either sender or receiver from denying a transmitted message. This service may be necessary for some applications.

- *Access control:* This service allows one to limit access to host systems and applications via communications links. This requires that the authentication service is available.

- *Key management and exchange:* This service allows for the negotiation of security keys between communicating entities. While the other security services can be implemented in a similar manner for unicast and multicast communications, the key management service is much harder to extend from unicast to multicast.

IPsec protocols are designed for providing authentication, integrity, confidentiality, and nonrepudiation. A limited protection against traffic flow analysis can also be achieved. These protocols can be used to secure IP over satellite links.

8.2.2   Wireless Wide Area Networks

8.2.2.1   GSM

GSM [20–23] is the pan-European standard for digital cellular communications. It was established in 1982 within the European Conference of Post and Telecommunication Administrations (CEPT). GSM provides enhanced features such as total mobility, high capacity, and optimal spectrum allocation, security, and services (e.g., voice communication, facsimile, voice mail, short message transmission, data transmission, and supplemental services such as call forwarding).

The security methods standardized for the GSM System make it the most secure cellular telecommunications standard currently available. Although the confidentiality of a call and anonymity of the GSM subscriber is only guaranteed on the radio channel, this is a major step in achieving end-to-end security. The subscriber's anonymity is ensured through the use of temporary identification numbers. The confidentiality of the communication itself on the radio link is performed by the application of encryption algorithms and frequency hopping, which could only be realized using digital systems and signaling.

Description of GSM Security Features

Security in GSM consists of the following aspects: subscriber identity authentication, subscriber identity confidentiality, signaling data confidentiality, and user data confidentiality. The subscriber is uniquely identified by the International Mobile Subscriber Identity (IMSI). This information, along with the individual subscriber authentication key (K_i), constitutes sensitive identification credentials. The design of the authentication and encryption schemes is such that this sensitive information is never transmitted over the radio channel. Rather, a challenge-response mechanism is used to perform authentication. The actual conversations are encrypted using a temporary, randomly generated ciphering key (K_c). The MS identifies itself by means of the Temporary Mobile Subscriber Identity (TMSI), which is issued by the network and may be changed periodically (i.e., during hand-offs) for additional security.

The security mechanisms are implemented in three different system elements: the subscriber identity module (SIM), the GSM handset or mobile equipment (ME), and the GSM network. SIM and ME together form the mobile station (MS). The SIM contains the IMSI, the individual subscriber authentication key (K_i), the ciphering key generating algorithm (A8), the authentication algorithm (A3), as well as a personal identification number (PIN). The GSM handset contains the ciphering algorithm (A5). The encryption algorithms (A3, A5, A8) are present in the GSM network as well. The authentication center (AuC), part of the operation and maintenance subsystem (OMS) of the GSM network, consists of a database of identification and authentication information for subscribers. This information consists of the IMSI, the TMSI, the Location

Area Identity (LAI), and the individual subscriber authentication key for each user. In order for the authentication and security mechanisms to function, all three elements (SIM, ME, and GSM network) are required. This distribution of security credentials and encryption algorithms provides an additional measure of security both in ensuring the privacy of cellular telephone conversations and in the prevention of cellular telephone fraud.

Figure 8.2 demonstrates the distribution of security information among the three system elements. Within the GSM network, the security information is further distributed among the AuC, the home location register (HLR), and the visitor location register (VLR). The AuC is responsible for generating the sets of random number (RAND), signed response (SRES), and K_c, which are stored in the HLR and VLR for subsequent use in the authentication and encryption processes.

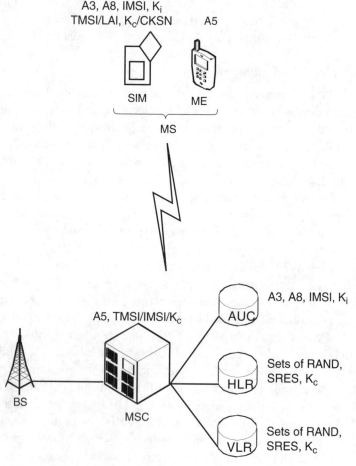

Figure 8.2 Flow of security information in GSM.

Authentication

The network authenticates the identity of the subscriber through the use of a challenge-response mechanism. A 128-bit RAND is sent to the MS. The MS computes the 32-bit SRES based on the encryption of the RAND with the authentication algorithm (A3) using the individual subscriber authentication key. Upon receiving the SRES from the subscriber, the network repeats the calculation to verify the identity of the subscriber. Note that the individual subscriber authentication key is never transmitted over the radio channel. It is present in the subscriber's SIM, as well as the AuC, HLR, and VLR databases. If the received SRES agrees with the calculated value, the ME has been successfully authenticated and may continue. If the values do not match, the connection is terminated and an authentication failure is sent to the MS. Figure 8.3 illustrates the authentication mechanism.

The calculation of the signed response is processed within the SIM. This provides enhanced security, because the confidential subscriber information such as the IMSI or the individual subscriber authentication key is never released from the SIM during the authentication process.

Signaling and Data Confidentiality

The SIM contains the ciphering key generating algorithm (A8), which is used to produce the 64-bit ciphering key. The ciphering key is computed by applying

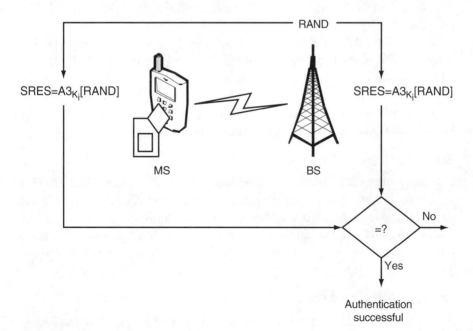

Figure 8.3 GSM authentication mechanism.

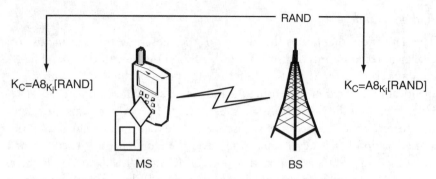

Figure 8.4 Ciphering key generation mechanism.

the same RAND used in the authentication process to the ciphering key generating algorithm (A8) with the individual subscriber authentication key. As will be shown in later sections, the ciphering key is used to encrypt and decrypt the data between the MS and BS. An additional level of security is provided by the ability to change the ciphering key, which makes the system more resistant to eavesdropping. The ciphering key maybe changed at regular intervals as required by network design and security considerations. Figure 8.4 shows the calculation of the ciphering key.

In a similar manner to the authentication process, the computation of the ciphering key takes place internally within the SIM. Therefore, sensitive information such as the individual subscriber authentication key is never revealed by the SIM.

Encrypted voice and data communications between the MS and the network is accomplished through use of the ciphering algorithm A5. Encrypted communication is initiated by a ciphering mode request command from the GSM network. Upon receipt of this command, the mobile station begins encryption and decryption of data using the ciphering algorithm and the ciphering key. Figure 8.5 demonstrates the encryption mechanism.

Subscriber Identity Confidentiality

To ensure subscriber identity confidentiality, the TMSI is used. The TMSI is sent to the mobile station after the authentication and encryption procedures have taken place. The mobile station responds by confirming reception of the TMSI. The TMSI is valid in the location area in which it was issued. For communications outside the location area, the LAI is necessary in addition to the TMSI. The TMSI allocation/reallocation process is shown in Figure 8.6.

Encryption Algorithms

A partial source code implementation of the GSM A5 algorithm was leaked to the Internet in June 1994. More recently there have been rumors that this

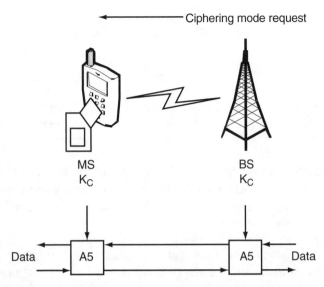

Figure 8.5 Ciphering mode initiation mechanism.

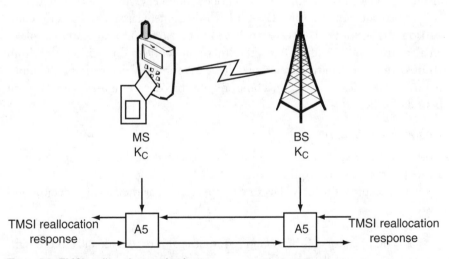

Figure 8.6 TMSI reallocation mechanism.

implementation was an early design and bears little resemblance to the A5 algorithm currently deployed. Nevertheless, insight into the underlying design theory can be gained by analyzing the available information. Some information on A5 is given below.

A5 is a stream cipher consisting of three shift registers, the sum of the degrees of these registers is 64, and they are initialized by the 64-bit session key and

Table 8.1
Brute-Force Key Search Times for Various Key Sizes

Key Length in Bits	32	40	56	64	128
Time Required to Test All Possible Keys	1.19 hours	12.7 days	2,291 years	584,542 years	10.8×10^{24} years

fed by the 22-bit TDMA frame number. A5 produces two 114-bit keystreams for each TDMA frame, which are XOR-ed with the uplink and downlink traffic channels. A5, although producing 114-bit keystreams, is rumored to have an effective key length of 40 bits.

Key Length. Key length is an important component of an encryption algorithm. Assuming that a brute-force search of every possible key is the most efficient method of cracking an encrypted message (a big assumption, mind you), Table 8.1 summarizes how long it would take to decrypt a message with a given key length, assuming a cracking machine capable of one million encryptions per second.

The time required for a 128-bit key is extremely large (as a basis for comparison, the age of the Universe is believed to be 1.6×10^{10} years. An example of an algorithm with a 128-bit key is the International Data Encryption Algorithm (IDEA). The key length may alternately be examined by determining the number of hypothetical cracking machines required to decrypt a message in a given period of time.

Limitations of Security

Existing cellular systems have a number of potential weaknesses that were considered in the security requirements for GSM.

The security for GSM has to be appropriate for the system operator and customer:

- The operators of the system require the ability to issue bills to the right people, and that the services cannot be compromised.
- The customer requires some privacy against traffic being overheard.

The countermeasures are designed:

- To make the radio path as secure as the fixed network, which implies anonymity and confidentiality to protect against eavesdropping;
- To have strong authentication, so as to protect the operator against billing fraud;

- To prevent operators from compromising each other's security, whether inadvertently or because of competitive pressures.

Beyond being cost-effective, the security processes must not:

- Significantly add to the delay of the initial call set up or subsequent communication;
- Increase the bandwidth of the channel;
- Allow for increased error rates, or error propagation;
- Add excessive complexity to the rest of the system.

The designs of an operator's GSM system must take into account the environment and much have the following secure procedures:

- The generation and distribution of keys;
- The exchange of information between operators;
- The confidentiality of the algorithms.

8.2.2.2 GPRS Security Versus GSM Security

In the General Packet Radio Service (GPRS) system, the frames are transmitted as cipher text from the MS to the serving GPRS support node (SGSN). This is done because the GPRS system uses multiple time slots in parallel in order to achieve a greater transmission rate. One GPRS phone can be allocated multiple time slots by the network, thus increasing the transmission rate of that MS. The frames can be sent in parallel time slots to the same BTS or to two different BTSs if the MS is handed over from one BTS to another.

To a BTS the use of one time slot is seen as a separate call. Thus, the BTS is unable to put the frames from different time slots together. This means that there has to be a network component that is able to receive the frames from one MS, defragment them, and send them onwards to the actual destination. The BTSs are also unable to decrypt the frames, because consecutive frames on one channel don not have consecutive frame numbers (see Figure 8.7). To simplify the implementation, the frames are decrypted at the SGSN where all of the frames end up, and it is thus easy to keep track of frame numbers. The solution is based on the ease of implementation and has not been implemented in order to increase system security. As a side effect, the GPRS system effectively prevents eavesdropping on the backbone between the BTS and SGSN, because the frames are still encrypted at this point. In GPRS, the triples from the HLR are transmitted to the SGSN and not to the MSC. Thus, security of GPRS depends largely on the placement and security of the SGSNs.

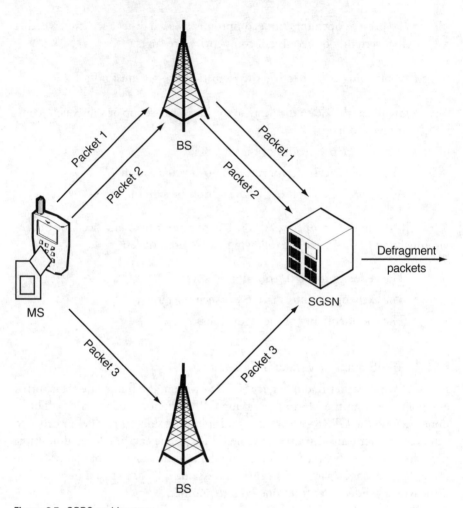

Figure 8.7 GPRS architecture.

The GPRS system uses a new A5 implementation as well, which is not known publicly. The fact that the frames are not decrypted at the BTS, but at the SGSN, rules out a couple of attacks. First, it is very hard to attack the A5 implementation when it is not known. Secondly, the K_c is not transmitted to the BTSs, and the transmission channel between the BTS and the SGSN is encrypted making it useless to monitor the backbone between the BTS and the SGSN. This does not mean that the GPRS security model would somehow be more secure than the GSM-only security model. It means that identical attacks do not work with GPRS that work with a GSM-only network.

As soon as the A5 implementation used in GPRS leaks out, the GPRS security model will be vulnerable to new attacks. It can be assumed that the implementation will leak out eventually or that the design will be successfully reverse engineered. As was stated above, the security of a crypto system should be based solely on the key. However, the majority of attacks against the GSM-only system are applicable against GPRS as well (e.g., the SIM-cloning attack). Additionally, the GPRS model introduces another security threat through the use of SGSNs, which know the triples from the HLR. This means that the security of the GPRS network depends largely on the positions of the SGSNs in the network architecture and the security of the SGSNs. If the SGSNs are vulnerable to an attack, then the triples are vulnerable as well.

Flaws

On April 13, 1998, the Smartcard Developer Association and the ISAAC security research group announced a flaw in the authentication codes found in digital GSM cell phones. This flaw allows an attacker with physical access to a target phone to make an exact duplicate (a clone) and to make fraudulent calls billed to the target user's account.

Three years later, there were indications that the GSM industry was taking steps to repair the security weaknesses in the GSM cryptographic algorithms. A patched version of COMP128 (COMP128-2) and COHP 128-3, are now available although they remain secret.

The GSM security model is broken on many levels and is thus vulnerable to numerous attacks targeted at different parts of an operator's network.

Assuming that the security algorithms were not broken, the GSM architecture would still be vulnerable to attacks targeting the operator's backbone network or HLR and to various social engineering scenarios in which the attacker bribes an employee of the operator.

Furthermore, the secretly designed security algorithms incorporated into the GSM system have been proven faulty. The A5 algorithm used for encrypting the over-the-air transmission channel is vulnerable against known-plain-text and divide-and-conquer attacks, and the intentionally reduced key space is small enough to make a brute-force attack feasible as well. The COMP128 algorithm used in most GSM networks as the A3/A8 algorithm has been proved faulty so that the secret key K_i can be reverse engineered over-the-air through a chosen challenge attack in approximately 10 hours.

All this means that if somebody wants to intercept a GSM call, he can do so. It cannot be assumed that the GSM security model provides any kind of security against a dedicated attacker. The required resources depend on the attack chosen. Thus, one should not rely solely on the GSM security model when transferring confidential data over the GSM network.

In addition to the possibility of call interception, the faulty COMP128 algorithm makes SIM cloning a threat as well, thus making it possible for an attacker to place calls at someone else's expense.

The reality is that although the GSM standard was supposed to correct the problems of phone fraud and call interception found in the analog mobile phone systems by using strong crypto for MS authentication and over-the-air traffic encryption, these promises were not kept. The current GSM standard and implementation enables both subscriber identity cloning and call interception. Although the implementation of cloning or call interception is a little bit more difficult, due to the digital technology that is used, compared to the analog counterparts, the threat is still very real, especially in cases where the transmitted data is valuable. GSM was an improvement over analog cellular in, say, 1992, but the improvements have been compromised, bringing the security level back to that of analog.

8.2.2.3 3GPP

3GPP [24–28] security was based on GSM security, with the following important changes:

- A change was made to defeat the false base station attack. The security mechanisms include a sequence number that ensures that the mobile can identify the network.
- Key lengths were increased to allow for the possibility of stronger algorithms for encryption and integrity.
- Mechanisms were included to support security within and between networks.
- Security is based within the switch rather than the base station, as in GSM. Therefore, links are protected between the base station and switch.
- Integrity mechanisms for the terminal identity (IMEI) have been designed in from the start, rather than that introduced late into GSM.
- The authentication algorithm has not been defined, but guidance on choice will be given.

When roaming between networks, such as between a GSM and 3GPP, only the level of protection supported by the smart card will apply. Therefore a GSM smart card will not be protected against the false base station attack when in a 3GPP network.

Many of the security enhancements required to 2G systems are intended to counteract attacks that were not perceived to be feasible in 2G systems. This includes attacks that are, or are perceived to be, possible now or very soon because intruders have access to more computational capabilities, new equipment has become available, and the physical security of certain network elements is questioned.

In order to perform the attacks, the intruder has to possess one or more of the following capabilities:

- *Eavesdropping:* This is the capability of intruders to eavesdrop signaling and data connections associated with other users. The required equipment is a modified MS.

- *Impersonation of a user:* This is the capability whereby the intruder sends signaling and/or user data to the network, in an attempt to make the network believe they originate from the target user. The required equipment is again a modified MS.

- *Impersonation of the network:* This is the capability whereby the intruder sends signaling and/or user data to the target user, in an attempt to make the target user believe they originate from a genuine network. The required equipment is modified BS.

- *Man-in-the-middle:* This is the capability whereby the intruder puts itself in between the target user and a genuine network and so has the ability to eavesdrop, modify, delete, reorder, replay, and spoof signaling and user data messages exchanged between the two parties. The required equipment is modified BS in conjunction with a modified MS.

- *Compromising authentication vectors in the network:* The intruder possesses a compromised authentication vector, which may include challenge/response pairs, cipher keys, and integrity keys. This data may have been obtained by compromising network nodes or by intercepting signaling messages on network links.

The first capability is the easiest to achieve; the capabilities are gradually more complex and require more investment by the attacker. Therefore, in general, an intruder having a certain capability is assumed also to have the capabilities positioned above that capability in the list. The first two capabilities were acknowledged in the design of 2G systems. 3G security, however, should thwart all five types of attacks.

8.2.3 Fixed Wireless Access

FWA typically has coverage between that of WLANs and cellular communications systems. This type of coverage requires high power transmissions. The main purpose of FWA is to provide network access to buildings through exterior antennas communicating with central radio base stations. In this way, users in a building are allowed to connect to the network with conventional in-building networks. In order to fulfill this requirement, the design of FWA accommodates for heterogeneous physical layers under the same MAC layer, providing different QoS. The IEEE standard associated with this kind of networks is the 802.16, published April 2002 [29].

8.2.3.1   IEEE 802.16

IEEE 802.16 addresses the first-mile/last-mile connection in FWA. It focuses on the efficient use of bandwidth between 10 and 66 GHz [the 2- to 11-GHz region with point-to-multipoint (PMP) and optional mesh topologies] and defines a MAC layer that supports multiple physical layer specifications customized for the frequency band of use.

The 10- to 66-GHz standard supports continuously varying traffic levels at many licensed frequencies for two-way communications. It enables interoperability among devices, so carriers can use products from multiple vendors, which warrants the availability of lower cost equipment.

Security Requirement

Security is implemented in the privacy sublayer, which is part of the MAC layer. The aim of the privacy sublayer is to provide:

- *Authentication:* This enables an entity involved in a communication to ensure the identity of the peer node it is communicating with.
- *Confidentiality:* This ensures that the information exchanged between entities is never disclosed to unauthorized recipients. Confidentiality is provided by means of encryption.
- *Integrity:* This guarantees that a message transferred is not corrupted by a third entity. A side effect of this is the provision of nonrepudiation, which ensures that the source of the message cannot deny having sent the message.
- *Secure key exchange:* Thanks to digital certificates and RSA public key cryptography, the key exchange phase is carried out in a secure way.

The Key Management Protocol

The first step to be performed when a subscriber station (SS) wants to transfer data is authentication. In 802.16, users must obtain access to the network and request to the BS an authorization key (AK) for message encryption. Moreover, the SS must periodically replace AKs with fresh ones, issuing reauthorization request messages to the BS. For this reason, a lifetime is associated to every key, after which a key is not valid anymore. To avoid service interruption, the SS always holds two AKs, and subsequent AKs have overlapping lifetimes. It is the SS's duty to ask the BS for a new AK when the older of the two held AKs expires.

In order to be authenticated, the SS presents an authorization request to the BS, containing a unique X.509 digital certificate, issued by the SS's manufacturer. Once the integrity of the certificate is verified [by means of the central authority's (CA) digital signature], the BS sends back to the SS an authorization reply, containing an AK, encrypted with the SS's public key, and the SS is automatically authorized to access the network.

This type of authentication is performed because of the existence of a CA that issues the certificate. The BS verifies that the certificate is valid by checking the CA's digital signature. This solution provides a very high security level against unauthorized access, but it can only be adopted in the presence of an infrastructure. For other situations (e.g., in cases of ad hoc networks, or generally whenever it is not possible to connect with an on-line trusted third party), this solution cannot be used.

The complete authentication procedure is carried out in the SS by an authentication finite state machine (FSM).

Two points are worth noting:

1. The SS must periodically send authorization requests to the BS in order to obtain fresh keying material. The events that can trigger a reauthorization cycle are:
   - *Authorization grace timeout:* The authorization grace timer times out. This timer, which is held by the SS, fires a configurable amount of time before the current authorization is supposed to expire, signaling to the SS to reauthorize before its authorization actually expires. If this is not accomplished, the SS shall be considered unauthorized and the access will be denied. The reason why AKs are dynamically changed is obvious. Since they are used for encryption, it is essential that their lifetime is smaller than the smallest time interval that an attacker needs to break the encryption algorithm. And in case of a successful attack, the dynamic changing of AKs limits the damages.
   - *Authorization invalid:* This event reflects either a failure in authenticating a key reply message sent by the BS, or a reception of an authorization invalid message sent by the BS. In both cases, BS and SS lose AK synchronization, and an interruption of the service can be expected, due to the lack of valid AK in the SS.
   - *Reauthorize:* The set of security mechanism implemented by the SS for a determined connection may vary dynamically. In this case, the SS shall be reauthorized.

2. The SS can fall in a silent state, in which the SS is not allowed to pass subscriber traffic. The transition to this state is triggered by some error

of permanent nature detected by the BS, as an unknown manufacturer of the SS's digital certificate, or an invalid signature on the certificate. This security measure prevents rogue SS from gaining access to the network.

Together with the AK, the BS also sends to the SS a set of Security Association Identities (SAID), which are descriptors of the SAs supported by the SS. An SA is basically a set of security mechanisms used over a certain connection. For each SA received, an SS starts a separate traffic encryption key (TEK) FSM (see Figure 8.8).

Keys Usage

Once the SS obtains the AK from the BS, every message exchanged between the two entities is authenticated by means of a HMAC with the SHA-1 hash algorithm. The download authentication key HMAC_KEY_D is used for authenticating messages in the downlink direction, while the authentication key HMAC_KEY_U is used for the same purpose in the uplink direction. The AK is sent from the BS to the SS in the authorization reply, encrypting it with the SS's public key. The protocol used is as follows. Public-key cryptography is used to establish a shared secret (i.e., an authorization key) between the SS and the BS. The AK is then used for creating a system of private keys for subsequent exchanges. The AK is also used for the generation of keys encryption keys (KEKs), which are used to securely exchange TEKs. The triple-DES (3-DES) algorithm is then used to encrypt TEKs. Since 3-DES keys are 64 bits long, and the result of the above truncation is a 128-bit key, one KEK is split into two parts, in order to provide two 3-DES keys. The 64 most significant bits of the KEK are used in the encrypt operation, while the remaining 64 are used in the decrypt operation. The two keys are obtained, denoted as k1 and k2, respectively; and the encryption and decryption are performed as follows: encryption: $C = E_{k1}[D_{k2}[E_{k1}[P]]]$; decryption: $P = D_{k1}[E_{k2}[D_{k1}[C]]]$. P is a 64-bit TEK plain text, and C is the result of this 3_DES encryption.

Flaws

The following aspects that are not covered by 802.16's security suite:

1. *The SS is authenticated, but the user is not.* Any theft of the SS would allow the thief to access the network and take advantage of the services provided.

2. *No mutual authentication is provided.* While the SS shall authenticate itself by means of a digital certificate, the BS is never required to authenticate with the SS. This allows a rogue BS to connect with a number of SSs and collect personal data.

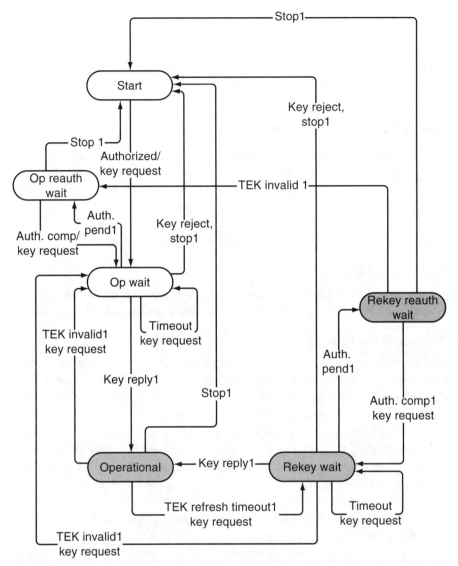

Figure 8.8 TEK state machine flow diagram.

3. *Privacy is not provided.* The authentication request that the SS sends to the BS is in the clear. This means that, even if a strong encryption algorithm secures the communication, the two identities of the parties involved in a communication are not protected. This problem could be simply solved by giving to the BS a private/public key pair and encrypting the authorization request with the BS's public key, but that still would not provide privacy. The encryption algorithm described

above is applied only to the MAC PDU, and not to the generic MAC header. Moreover, all MAC management messages are sent in the clear to facilitate registration. So an eavesdropper could obtain the identities of the two parties involved in a communication by simply eavesdropping any of the packet transferred and reading the MAC header.

8.2.4   Wireless Local Area Networks

WLANs mostly operate using either radio or infrared techniques. Each approach has its own attributes that satisfy different connectivity requirements. The majority of these devices are capable of transmitting information up to several 100 meters in an open environment. WLANs components consist of a wireless network interface card (STA) and a wireless bridge (AP). The AP interfaces the wireless network with the wired network (e.g., Ethernet LAN).

8.2.4.1   IEEE 802.11

WLANs based on IEEE 802.11 [30–38] were the first and most prominent in the field. IEEE 802.11 has different physical layers working in 2.4 and 5 GHz.

Wired Equivalent Privacy

The principal aim of the security suite implemented in 802.11 is to address the need for authentication and confidentiality. The protocol is based on a symmetric key, shared between the AP and the STA. The issue of securing the key exchange is not taken into account in the standard.

For authentication, two modes are specified: open systems authentication (OSA) and shared key authentication. The former basically means no authentication: a STA requesting access to the network does not need to carry out any additional operation in order to obtain it. In the second authentication mode, the AP uses a typical challenge-response mechanism, based on the symmetric key, in order to verify the STA's identity. This process is shown in Figure 8.9.

When an AP receives a registration request from a STA, it sends to the STA a random number internally generated, which represent the challenge text. The STA then encrypts this random number with the symmetric key, and sends it back to the AP. The AP performs the same encryption operation, and verifies that the result of the operation is identical to that sent by the STA. If so, the AP has the proof that the STA holds the symmetric key, and the authentication is successful. The wired equivalent privacy (WEP) algorithm is used for packet encryption, as well as for encryption of data packets after authentication. The aim of this algorithm is to provide WLAN with the same security level existing in wired LAN. WEP uses the RC4 algorithm based on a 40-bit symmetric key and a 24-bit initialization vector (IV). WEP was later enhanced to work with a 104-bit key and 24-bit IV. An integrity check value (ICV) is calculated in order

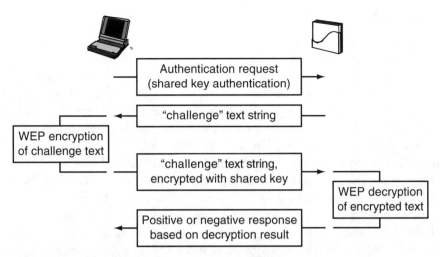

Figure 8.9 Authentication process in the shared key authentication mode.

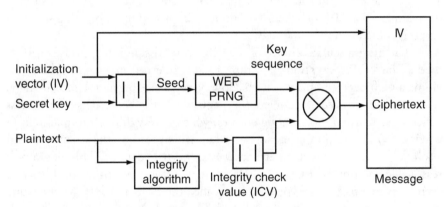

Figure 8.10 WEP cipher block diagram.

to provide integrity. The block diagram of a WEP cipher is depicted in Figure 8.10.

The message in plaintext is concatenated (the symbol "||" in Figure 8.10) with the ICV (CRC-32), and then XOR-ed with the encryption key. This is generated by a pseudo-random number generator (PRNG), and the seed at the input is the concatenation (the symbol "||" in Figure 8.10) of an IV with the symmetric key. At the output of the cipher, the ciphertext is concatenated to the IV. In this way, the receiver easily decrypts the message by reading the value of the IV and performing the same operations as in transmission, but in inverse order.

WEP Flaws

In the IEEE 802.11 standard only 5 pages out of more than 500 are dedicated to security. Instead of trying to find a patch partially improve it, the easiest and most effective solution would be to create a new security protocol from scratch. Unfortunately, this is not possible, due to the fact that nowadays millions of 802.11 devices implementing the WEP algorithm have already been shipped and deployed. For this reason, the only way to improve upon security is through software upgrades, but this solution is constrained by the fact that operational devices today have very low computational capabilities in order to remain cost-effective. For example, in order to strengthen the confidentiality provided by the WEP algorithm, a 3-DES algorithm can be used, as in 802.16 networks. Now, implementing 3-DES with a popular C language, the cost of an encryption is about 180 instructions per byte. With the actual 802.11b throughput, which is around 7 Mbps, the encryption requires about 157.5 million instructions per second in order to be performed. For this function be carried out on a 20-MHz processor (which typically the AP are equipped with), it would roughly require 10 times the maximum number of instruction supported. It can be easily deduced, then, that 802.11's security cannot be improved upon by using standard cryptographic methods.

The most discussed choice of the IEEE commission is the misuse of RC4, used as the WEP encryption algorithm. RC4 is an excellent stream cipher, providing a high degree of privacy with a low performance penalty. Unfortunately, it is inappropriate to datagram environments like WEP, because its effectiveness is tied to synchronization among entities, which is very difficult to maintain in wireless communication where loss of data is widespread. In the attempt to find a solution to this problem, a per-packet RC4 key was defined. This would be a reasonable solution to RC4 loopholes, but it introduces further problems. It even leads to a particular type of attack, called the FMS attack (from the name of the investigators who discovered it, S. Fluhrer, I. Mantin, and A. Shamir), which is easy to perform because the IV, which constitutes the seed for the key stream generator, is transferred in the clear over the medium, exposing the key stream to direct attack.

The configuration of the base key (i.e., the symmetric key) used in the WEP algorithm is manual. For this reason, it is often kept constant for a long period. This drastically reduces the key space available by the WEP cipher. The IV is 24 bits long, which means that at most 2^{24} different keys can be associated to the same base key. With an average of 1,800 data messages per second supported by 802.11b, this number of keys is sufficient for about 2.5 hours. If the base key is not replaced after this time span, as often happens, key collision is an ordinary event. But as if this was not enough, the 802.11 standard does not state any rule for IV refresh. As a result, many real-life implementations operate with a fixed IV and employ the same RC4 key to encrypt every packet. This way, the RC4 key is kept to a constant.

The ICV should guarantee integrity, but CRC-32 does not fulfill this requirement at all. The problem is that the CRC-32 construction and XOR-based encryption commute. Thanks to this, an attacker can generate a valid unencrypted packet and the corresponding ICV, and XOR them with the encrypted packet and the original ICV, and still obtaining a valid packet.

WEP provides no replay protection. An active attacker can record a message and retransmit it in another moment, with or without manipulating it. Moreover, both the destination and the source addresses are neither encrypted nor protected by the ICV. This allows an attacker to intercept a packet, change the destination, and retransmit it to a station that is not authorized to participate in the communication. The AP receives the packet and considers it genuine, consequently decrypting it and forwarding it to the unauthorized station. Forgery attacks are therefore very common, because they are very easy to perform without any need for high-tech devices.

IEEE 802.11i

IEEE 802.11i is the security enhancement standard issued in July 2004. Until IEEE 802.11i standard was available, IEEE 802.11 vendors provided some security solutions to bridge the gap. These solutions started with extended WEP key size, which was adopted by the standard, as well as providing RADIUS and MAC address–based authentication, IEEE 802.1x port–based user authentication, and AES-based encryption.

Seeing the market situation, Wi-Fi (the IEEE 802.11 interoperability industry alliance) is introducing the Temporal Key Integrity Protocol (TKIP) as a simple but secure intermediary solution. This solution is usually known as Wi-Fi Protected Access (WPA) and is available from most vendors. WPA provides enhanced data encryption through the TKIP, user authentication via 802.1x and EAP, mutual authentication, and, for ease of transition, Wi-Fi certified products are software upgradeable.

802.1X

802.1X [39] is the IEEE's answer to 802.11's incontestable loopholes. It is a standard for port-based network access control, and offers an effective framework for authenticating and controlling user traffic, and for keys periodical refresh. It makes use of the Extensive Authentication Protocol (EAP), which was not expressly created for wireless networks but which fits well into 802.11. The authentication algorithm actually used is left open; 802.1X supports multiple authentication methods, such as token card, Kerberos, one-time passwords, certificates, and public key authentication.

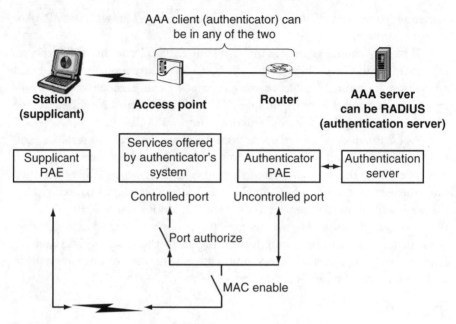

Figure 8.11 Dual port operation in 802.1X. (PAE: port access entity.)

The basic idea is that for each supplicant asking to access a service, two ports are held in the authenticator. One port is uncontrolled, and it is used only for initial EAP messages, while the other port grants access to the authenticators' services, and it is open only after user authentication. A block diagram of the system is shown in Figure 8.11.

In order to accomplish the authentication procedure, the following steps are performed:

1. The supplicant (the STA, in the case of 802.11) sends an EAP-start message to the authenticator through the uncontrolled port.

2. The authenticator (the AP, in the case of 802.11) replies to the supplicant with an EAP-request identity message.

3. The supplicant sends to the authenticator an EAP-response, carrying the formers' identity.

4. The authenticator interrogates an authentication server, which uses an algorithm (open to different implementations) to verify the supplicant's identity.

5. The authentication server sends to the authenticator either an accept or a reject message, which means that the supplicant is either authenticated by the server or not.

6. In case of successful authentication in the authentication server, the authenticator sends to the supplicant an EAP-success message and opens the controlled port to the supplicant. If the authenticator receives a reject message from the authentication server, the message is forwarded to the supplicant, and the access to the controlled port is denied. In this case, for each access denial, the authenticator shall augment the time between two subsequent requests from the supplicant, as a countermeasure against denial of service (DoS) attacks.

Note that this implementation of 802.1X provides authentication regardless of the use of WEP cryptography. On the other hand, the drawback of one-way authentication remains. The port on the client is always left uncontrolled—that is, the server is assumed to be a trusted entity, which is not necessarily the case, especially in the wireless environment.

The authentication server provides a series of advantages. The first one is that it makes impossible any man-in-the-middle type attack. Another great advantage is that it is easy to think about a key exchange protocol relying on the authentication server. An EAP type mechanism for key generation and exchange is known as transport layer security (TLS). It is based upon a public keys mechanism, which is used to securely transmit session keys from the authentication server to both the supplicant and to the server. To make the algorithm stronger, the mentioned session keys can be used as keys encrypting other keys periodically generated by the authenticator. This way, the refresh of WEP keys would not be a problem anymore.

This provides an optional capability to improve on WEP security, called broadcast key rotation (BKR). When BKR is turned on, the AP periodically broadcasts new keys to the STAs, encrypting them with the current WEP key. These keys are then used as base keys for WEP key generation. In this way, the key space is considerably increased, and the probability of key collisions lowers. Note that this mechanism can be implemented regardless of the presence of any infrastructure, which is saying, in ad hoc 802.11 networks. The EAP-based authentication protocol, however, needs the aid of an infrastructure.

Another optional function in 802.1X to solve WEP's loopholes is called TKIP. TKIP was developed by IEEE 802.1X TGi (Task Group "i") and proposed as a short-term patch for 802.11 security. TKIP was then kept as an option in the long-term patch, which is 802.1X. The principles that TKIP uses to improve on WEP security are as follows:

1. It establishes a key hierarchy. At the higher level of the hierarchy there is a master key, which is shared between a STA and an AP, and which is established as a side effect of authentication. The master key is used

to encrypt key encryption keys, which are generated either by the AP or by a server (in this case they are transmitted to the AP by means of a public keys system) and represent the intermediate level of the hierarchy. These keys are then used to encrypt temporal keys, which lie on the lowest level of the hierarchy, and are often replaced. Thanks to this hierarchy, master keys are rarely used, and no key is used for long periods, considerably reducing the number of possible attacks and the probability for them to be successful.

2. In addition to the classical ICV, for each MSDU another integrity check is performed, called MIC (standing for Michael, but actually meaning message authentication code), a 64-bit code, obtained from the message and the current temporal key. With no doubt, the MIC constitutes a deterrent for forgery attacks. TGi recognized, however, that the level of protection that it provides is too weak to afford much benefit by itself. That is why TKIP complements MIC with countermeasures. In fact, upon receiving a series of packets, if the MIC check fails two subsequent times, the receiver assumes that it is under attack. In this case, keys are deleted and the communication is interrupted. It is evident that this is another patch, which partially solves the problem of forgeries, but introduces other problems related to DoS attacks.

3. TKIP makes more complex the generation of the WEP keys, which is one of the biggest problems of WEP security. This is accomplished by means of two key mixing. In the first mixing phase, the temporal key is combined with the TA, in order to obtain different keys for the STAs, even if they start from the same temporal key. The result of this combination is an intermediate key. Note that the intermediate key needs to be generated only once every temporal key replacement. In the second mixing phase, the intermediate key is combined with the MSDU fragment to be transmitted, so generating a per-packet key. This key is then interpreted by the classical WEP encapsulation as the concatenation of a base key plus an IV. In this way, the goal of the second key mixing is reached, which is to decorrelate the IV and the per-packet key.

All the operations performed by TKIP do not need the deployment of new hardware. From this point of view, its implementation can be considered cost-effective. On the other hand, TKIP significantly degrades the performance of APs, by occupying every spare CPU cycle. Since TKIP is an option in 802.1X, the loss of performance with it is unfortunately very often a good reason to keep it off.

8.2.5 Wireless Personal Area Networks

The 802.15 WPAN [38, 40] effort focuses on the development of consensus standards for PAN or short-distance wireless networks. These WPANs address wireless networking of portable and mobile computing devices such as PCs, PDAs, peripherals, cell phones, pagers, and consumer electronics; allowing these devices to communicate and interoperate with one another.

8.2.5.1 IEEE 802.15

The IEEE 802.15.1 has derived a WPAN standard based on the Bluetooth v1.1 Foundation Specifications.

Keys and Parameters

The parameters involved in the security algorithms are as follows:

- A Bluetooth address (BD_ADDR), which is a 48-bit IEEE address, given to the device by the manufacturer.
- A 128-bit private key used for authentication, which is generated during the initialization procedure and remains unaltered during the whole presence of a user in a piconet.
- A private key used for encryption, ranging from 8 to 128 bits, which is generated when the link manager issues a LMP_encryption_mode_required message, which can happen in any moment of a data transfer. The size of the encryption key is configurable in order to cope with the different requirements for encryption algorithms in different countries.
- A 128-bit RAND, which is derived within a Bluetooth unit. The principal requirement for this number is that it has to be nonrepeating and randomly generated.
- A four-decimal digit PIN, which the user has to introduce before initialization.

Five types of keys can be generated for different purposes:

1. Unit key (K_A), internally generated within the device A;
2. Combination key (K_{AB}), generated by two entities A and B;
3. Temporary key (K_{MASTER}), generated by the master and shared in the whole piconet;
4. Initialization key (K_{INIT}), which is the first key to be generated and from which the others derive (with the exception of the master key);
5. Encryption key (K_C), generated within the master and one or more slaves.

Key Generation

The first step to be performed in the initialization phase is the generation of a 128-bit initialization key, K_{INIT}. The PIN has to be introduced by the user in both devices involved in the initialization process. It can also happen that a device has a fixed PIN (e.g., in case of devices with no man-to-machine interface), but two devices with fixed PINs cannot be paired. The PIN is then augmented with the BD_ADDR of the device, to form a 128-bit PIN'. In case one of the two units has a fixed PIN, it shall use the other unit's BD_ADDR to augment it. This way, it is ensured that the K_{INIT} depends on the entity of the unit with variable PIN, which is considered more reliable, since the PIN was introduced by the user. It does not have to be protected, since its only aim is to hide to the user the K_{INIT} and to ensure that the introduction of the same PIN does not generate the same K_{INIT}.

Next, a link key is generated to securely exchange data and perform authentication between two entities. It can be a unit key as well as a combination key. A unit key is generated in a single unit when the Bluetooth unit is in operations for the first time. The key is then sent to the other unit simply by XORing it with the initialization key. The generation of a combination key provides higher security than the preceding case, because the key is dependent on two units. The generation process is a bit more complex.

The choice between a unit key and a combination key is based on the level of security required by the application and the storage capacity of the units involved. In fact, if a combination key is used for every link, the number of keys to be stored for every device is proportional to the number of devices in the piconet.

Encryption is optional in Bluetooth, and it is activated by a specific LM command generated within the master, potentially long after the authentication phase. The encryption key is derived by the so-called E_3 algorithm. Note that the result of the E_3 algorithm is a 128-bit key, but the key can subsequently be cut in order to obtain an encryption key of the length previously negotiated between the master and the slave.

For broadcast communications, a master is theoretically capable of handling different K_cs for every slave, but this may lead to a loss of capacity. Moreover, a slave may not be capable of handling different K_cs for broadcast and individually addressed messages. For this reason, the master can use a temporary key, the master key, replacing all the link keys for every slave. From the master key (K_{MASTER}, 128 bits) a common encryption key will then be generated, used for both broadcast and individually addressed traffic. The master key is generated through the same algorithm used for the generation of the initialization key (E_{22}).

After K_{MASTER} is established, a common encryption key can be used in all the devices of a piconet.

Keys Usage

Link keys are either temporary or semipermanent. A temporary link key has a lifetime equal to a session, which means that once this has expired, a new link key has to be generated. Typically, temporary keys are used in point-to-multipoint configuration, as in the case of the master key, for practical reasons of limited storage or network capacity. In other cases, semipermanent key are established, such as combination or unit keys. These keys are stored in a nonvolatile memory and can be reused in several subsequent connections, straightforwardly performing the authentication and skipping the initialization phase. This type of key is not permanent because it is possible to change or delete them. This necessity may be due to either practical reasons, such as limited memory, or security reasons, such as the barring of a user previously considered trusted. Note that the procedure for changing a link key can be straightforward in case of combination keys and rather complex in case of unit keys. The change of a combination key can be performed by creating a new one using the old combination key as the link key; while the replacement of a unit key requires the reinitialization of all units trying to connect. It can therefore be deduced that, in case of unit keys it is advisable not to replace them, and in case of combination keys this can be a good policy to enforce the security of the system. However, no mechanism for key replacing is specified in the standard.

Authentication. After the generation of a link key (either a combination key or a unit key), a mutual authentication is performed to confirm the success of the transaction. The scheme on which authentication is based is a typical challenge-response mechanism.

The verifier issues a random number and transmits it to the claimant in the clear. The claimant encrypts this number with the link key and sends the result to the verifier, which computes the same operation and verifies the correspondence of the two results. Note that the claimant's BD_ADDR is encrypted together with the random number, in order to avoid a possible reflection attack. Also, the authentication can be mutually performed if the application requires. This means that the verifier can be either a master or a slave. As a side effect of authentication, a number called the authentication ciphering offset (ACO) is generated, which is needed for the generation of encryption keys.

It is worth noticing that the standard also recommends a countermeasure against DoS attacks. Imagine that a hacker aims at jamming the network by continuously sending authentication requests with different keys. The countermeasure taken by Bluetooth, in case an authentication procedure fails, is to establish a waiting interval before a new authentication request is accepted from the same unit. This interval is increased each time that a subsequent failure is detected, in an exponential fashion. Once the authentication is successful, the interval decreases. The only drawback of this mechanism is that, in order to be effectively performed, every unit should keep a list of waiting intervals, corresponding to the units it has established contact with. In practical implementation, this list has a finite length, with the entries being replaced according to a FIFO policy.

Encryption. For encryption, a stream cipher algorithm is used, encrypting separately each payload. The cipher algorithm E_0 calculates the key stream as: $K_{CIPHER} = E_0(BD\_ADDR, CLK_{MASTER}, K_C)$, where BD_ADDR is the master address, CLK_{MASTER} is the clock of the master, and K_C is the encryption key previously calculated from the random number EN_RAND, sent to the slave by the master. The clock used for encryption is updated at every time slot, so that a new key is generated at every packet. For packets covering more than one slot, the key is derived from the value of the clock in the first time slot occupied by the packet.

Once the key stream is calculated, it is directly added modulo two with the data, on transmission for the encryption and on reception for the decryption. The E_0 algorithm can be conceptually divided into three parts. The first one is the payload key generator, and it simply orders the four inputs to the algorithm and shifts them into four linear feedback shift registers (LFSRs). The second part of the algorithm is the key stream generator, which represent the encryption engine of the algorithm and contains combinational and blend logic. The third part is the summation modulo two between the key stream and the plaintext packet.

Flaws

The problem with Bluetooth security is that the security suite was designed with the aim of providing a reasonable, but not high, security level. This was done under the wrong conviction that short-range coverage, together with low data rate, is synonymous of low security requirements. But as Bluetooth technology has spread in the past years, it has also demonstrated its capability of covering many more types of services than it was expected, with a subsequent need for higher security. Unfortunately, the constraints imposed by device features remain the same, for which it is not possible to cope with security problems simply by increasing and strengthening security mechanisms. Therefore, security must be enhanced in Bluetooth only in those cases in which services actually require a higher security level, and the solution cannot be the outcome of stringent trade-offs.

The Tacky PIN. The PIN code represents the first problem, for two main reasons. First, the security of the system depends on the link key; the security of the link key depends on the initialization key; and the security of this depends on the PIN. It can therefore be concluded that the overall security of Bluetooth is as strong as the PIN. Now, consider that the PIN is generally a four-digit number, for which there are just 10,000 possible combinations. Adding the fact that 50% of PINs used is "0000", one can conclude that a brutal attack to break the PIN is the most likely to be performed, simply for its ease of realization. In fact, the mechanism of an increasing waiting interval as a countermeasure against DoS or PIN-directed attacks through repeated attempts is feeble and memory wasting. From this point of view, it would be sufficient to increase the number of digits of the PIN code to proportionally increase the security of the system. But here comes the second problem with the PIN, which is its usability. It is obvious that the longer the PIN, the tackier the pairing process, since the introduction of the PIN is manual. In case there are many devices in a piconet, then, and each one has to be initialized separately, the problem gets insurmountable. Moreover, a PIN number can be directly caught by an attacker, independently of its length, if the pairing procedure is performed in a nonsecure place—that is, the attacker can physically spy on the user while he introduces the PIN. The only possible solution to this is application-level key agreement software with long PIN codes (up to 16 octets), so that there is no need to enter the PIN code physically, but it is exchanged, for example, by means of a Diffie-Hellman key agreement protocol, even if this introduces further complexity and the risk of impersonation attacks.

The Unit Key. The use of a unit key as a link key represents another problem for Bluetooth security. Suppose that A and B use A's unit key as a link key, and that C is subsequently paired with A with the same unit key. In this scenario, the communications between A and B, and A and C can be disclosed by any of the three units. This is a clear breach in confidentially. Moreover, C can also calculate the encryption key used by A and B (since their addresses are public) or even authenticate with A faking to be B. This problem is so evident that even the Bluetooth SIG recommends the use of unit keys as link keys as little as possible.

Lack of Privacy. Bluetooth does not even consider the issue of privacy. The Bluetooth device address is a 48-bit IEEE address that is unique for each unit. This address is publicly known, and it can be obtained either by inquiring the unit via the man-to-machine interface (MMI), if present, or via an inquiry routine by means of another Bluetooth unit. Moreover, during data exchange, the Bluetooth address is often transmitted in the clear [e.g., in frequency hopping synchronization (FHS) packets]. Even if encryption is activated, the packet header is left in plaintext. Considering that a Bluetooth device is generally a personal object. This means that the exact position of the user can be easily derived by eavesdropping. Unfortunately, dynamic changes to the Bluetooth address are not advisable because, even if this would solve the problem of privacy, it would introduce other problems, such as the uncertainty on the uniqueness of a

Bluetooth address. Moreover, this solution would impact the authentication procedure, which relies on the fact that an address does not ever change. So, if this solution were adopted, two Bluetooth devices wishing to communicate should be time paired, even if they have already communicated in a previous time.

Authentication. The issue of authentication is the most problematic in 802.15 networks. Let us deduce its weakness from some possible types of attack. First of all, since no time stamp is applied to transmitted packets, a *replay attack* can be performed by every eavesdropper, even if he or she ignores the contents. In other words, upon receiving a packet, there is no means to verify its freshness. A simple solution involves attaching a time stamp to every sent packet, so that the receiver can consider a packet expired upon receiving if a determined time lapse has passed with respect to the time stamp attached to the packet. An equivalent solution can involve attaching a sequence number to every sent packet, so that out-of-sequence packets are automatically discarded. Note that this solution can be effectively adopted only if packet forgery is prevented. Nevertheless, the 802.15 standard does not specify any usage of MAC codes for such aim.

An active attacker can really let his imagination run away with the possible *forgery attacks* he can perform over any packet. Since just a simple CRC is calculated over the packet header, the attacker can change its parameters as he likes and subsequently generate the correct parity check to be associated to the header. For example, the 3-bit active member address (AM_ADDR) contained in the header can be changed, without the possibility of detecting the forgery upon the receiver. When encryption is activated, this type of attack is prevented, since the encryption key (K_c) is derived from the permanent BD_ADDR of a device, which is associated with the AM_ADDR of the same device. So, if the AM_ADDR field in the header is changed, the decryption process will give nonsense as result, so that way the forgery is detected. Therefore, the way to prevent forgery attacks is either to encrypt every packet or to introduce a MAC code. It is obvious that both solutions involve the use of CPU cycles, with consequent increase of latency, and for the second solution, additional overhead. Again, what needs to be done is the outcome of a delicate trade-off.

Since the authentication phase is not tied to the data exchange, the hacker may correctly think that if he somehow breaks the authentication procedure, he will then be free of using the services provided by the master, since he will be considered trusted. This is precisely what happens in the *man-in-the-middle attack*. Let us describe this attack in case a combination key used as a link key (which is the worst case for the attacker). Imagine that A and B are not connected, and C wants to perform the attack, connecting to A and faking to be B. The master A will send to C a challenge for the authentication. C will establish a connection with B, sending it the same challenge. B will correctly answer to the challenge to C, and C will simply forward the result to A, who will successfully perform the control over it. The unit C is therefore authenticated and considered trusted by A, who will offer it the services restricted to B. As long as A does not require encryption, C will be able to masquerade as B and make use of its privileges, even if it never knows the link key between A and B.

What makes this attack possible is that in an ad hoc network there is no way to assure that a message is associated with the address contained in the message itself. In other words, whenever a third trusted party cannot be invoked (such as a certification authority), other solutions have to be contrived. A possible solution is to include the frequency hoping sequence in the hash function that A initially computes, instead of transmitting it in the clear, so that C cannot read it without knowing the link key between A and B, and therefore cannot physically communicate with A. An alternative solution is that A sets a timer after sending the challenge. Because Bluetooth coverage radius is small and each device is allowed to transmit only in precise time slots, the round trip delay is guaranteed. A man-in-the middle attack needs additional time slots to be performed, so the timer in unit A can be easily set to a value that prevents the interposing of other entities on the way to B.

8.3 Applications and Services

Based on the above discussions, three major application scenarios of heterogeneous networks are described in this section.

8.3.1 Low Data Rate

Low data rate (LDR) networks represent a new paradigm for reliable environment monitoring and information collection. They hold the promise of revolutionizing sensing in a wide range of application domains because of their reliability, accuracy, flexibility, cost-effectiveness, and ease of deployment. Furthermore, in future smart environments, it is likely that LDR networks will play a key role in sensing, collecting, and disseminating information about the environment.

Chapter 2 discusses the applications, services, and requirements of LDR networks, and then proposes an adaptive security architecture with crypto algorithm/protocols, which provides authentication and confidentiality under the constraints of code size, CPU, and memory size.

8.3.2 Ad Hoc Networking

In Latin, ad hoc means "for this"; further it means "for this purpose only." A wireless ad hoc network is an infrastructure-less collection of communication devices that wish to communicate with available links that have no predetermined organization. Each node dynamically determines the responsibility of direct communication. Each node has a wireless access interface (Bluetooth, WLAN, Hiper-LAN/2, UWB) and is free to enter or leave the network at any time. Due to the limited range of a node's wireless interface, multiple hops may be needed for communication. Ad hoc networks can function as stand-alone networks by meeting direct communication needs of their users, or as an addition to infrastructure-based networks to extend or enhance their coverage. This kind of communication becomes a viable solution especially in situations of missing or incomplete networks. Applications of ad-hoc communication include sensor networks, commercial and educational use, emergency cases, and military communication.

Chapter 7 gives a qualitative comparison of the routing protocols and proposes an adaptive secure routing protocol for ad hoc networks. It also proposes a light-weight AAA infrastructure for ad hoc networks to solve the need of authentication, authorization, and accounting.

8.3.3   Mobility

To accommodate the heterogeneous nature of devices and networks that might be involved in heterogeneous environments, this AAA infrastructure needs be light weight and able to adapt itself to the needs of the users and applications, as well as to the capabilities of the devices and networks in terms of the available bandwidth and processing capabilities.

AAA mechanisms are usually designed for the authentication and authorization of single users and devices. Heterogeneous networks usually consist of a number of interconnected devices that might move from one access network to another and require to be reauthenticated and reauthorized. In addition, heterogeneous networks and the applications running on them need to determine the security resources and services that are provided by their AP upon initiation and negotiate their security level. Mobile heterogeneous networks need to maintain their current security level, which is required by the end user and services as they move from one AP to another or renegotiate new security levels and services if they are not available from the new AP.

Furthermore, heterogeneous networks need to negotiate security levels with the other heterogeneous networks they want to communicate with, and also with the core network to ensure that the agreed-upon security level can be provided end to end. The agreed-upon end-to-end security level must be provided and maintained by both the ad hoc and infrastructure-based segments of the heterogeneous networks.

Chapter 4 investigates the support of services and applications on top of heterogeneous networks in a private and secure manner. Adaptive security mechanisms and light-weight AAA infrastructures are proposed that provide continuous, on-demand, end-to-end security in heterogeneous networks.

8.4 Security Requirements

Today, security has become one of the most noticeable issues in networks. Table 8.2 presents general security issues of low data rate, ad hoc, and mobility architectures. The security goals are: data confidentiality, data integrity, authorized access to data and services, availability and correct functioning of services, authenticity of communicating entities, and accountability to actions performed or caused by an entity.

Table 8.2
Security Requirements for Heterogeneous Networks

Confidentiality	All transferred data in foreign networks must be protected to the same or higher level of confidentiality as in the home network.
	All mobility management information must be protected from any unauthorized disclosure, whether in transit or storage at any point in the network.
	The operators must protect packets they forward on behalf of their users by encrypting this traffic during exchanges with other operators.
Integrity	All transferred data and databases must be protected against any unauthorized modification or deletion. This is particularly important when exchanging information related to users' accounts and locations. It must also be possible to prove the integrity of data for nonrepudiation purposes.
	Consequently, the integrity of user data and control signaling, such as AAA data between AAA infrastructures elements and mobility management data (e.g., binding update) between the mobile, its home network, and its correspondents, should be provided.
Authentication	The network authenticates the user and/or its mobile node before granting access service.
	The user/mobile node authenticates network.
	Authentication will occur between network elements in order to set up/maintain security associations.
	Each network would usually authenticate all information (packets) before allowing these to enter its internal network, in order to protect this internal network and its users.
Authenticity	Operators must make sure that all its stored (management) information is authentic.
	Operators must be able to verify the authenticity of users requiring their services, perhaps without authenticating the users directly (in fast handover it is not necessary to authenticate the mobile nodes again after handover without changing network operator because information from old FA or base station can be used).

(*continued*)

Availability	All information, resources and network services will be protected against DoS attacks. This is the basic requirement of providing service to end users according to a given policy defined in the contract.
	Moreover, a malicious MN (or an MN that has failed to authenticate) should not be able to flood the mobility management equipment of the network operator with malicious packets (e.g., attempt to create large numbers of sessions).
Authorized Access	Operators must ensure that users are authorized to access the services that they require, either by the operator itself when the user is in home network or by user's home network or trusted third party when it is a roaming user.
	Operators must ensure that all claims to access confidential data are requested by those authorized to do so before granting access to this data.
Accounting	Accountability of usage of resource.
	Accountability of user communications.
Location and Identity Privacy	Information of user location and identity in control signaling (e.g., care of address of a mobile node, user identity in paging signaling, and authentication process) must be made transparent to those having a legitimate interest in the information.
	Those who cannot prove a legitimate interest for information of a user's location must not be able to derive this information.
	Signaling over the air interface, which includes the identity of a user (e.g., implicitly by reference to a MN's home address), should be encrypted, or a temporary identity should be used to prevent air interface monitoring being used to identify users.

However, it is vitally important that these capabilities are designed from the start, as they will have an impact on the system requirements. Business cases should show the effect of fraud and the costs of protection.

8.5 Research Challenges

Networks that are known and used today are based on the preestablished relationships between a network subscriber and the network operator. Before a user can access a network and its services, he has to establish the appropriate contract with the network operator. This contract, along with the financial details, defines all technical details required for the proper configuration of connections and the user's device setup. For different networks belonging to different stakeholders, different setups are required. The level of technical expertise required for a particular setup may range from a very basic knowledge to high-level technical expertise. Since one tends to regularly use only a limited set of networks (mobile network, Internet provider, office network), users are usually well aware of where and how to use them. However, as soon as one moves out of their everyday environment various problems arise: the ISP access number is not known and hard to find, roaming is not possible due to the lack of agreement between network operators, IP address or domain address required for access to the office network is not appropriate, and so on. Obviously, many would benefit if solutions for automatic network and service discovery and connection establishment were provided.

At the same time, various devices with different wireless communication interfaces are proliferating. Mobile phones, laptops, PDAs, pagers, game consoles, and other similar devices are becoming an integral part of everyday life of many people. There is a growing need for an easy, secure, and simple establishment of connections between these devices to support data exchange in various scenarios: people attending a meeting or conference want to exchange presentations, those traveling together on a bus or train want to play games together, and so forth. The variety of wireless devices will enable deployment of different networks that will offer many interesting, location- or event-based services: wireless sensors will provide information about traffic in the area or will guide drivers to the nearest free parking space, shopping centers will inform customers about special offers, various tourist information will be available across cities, and each monument will be able to "tell" its history. We will securely access and use these services opportunistically, depending on our current needs, location, and time, which would be independent of underline access technology. Since a huge number of service providers and networks will be active in various locations and in different time periods offering different services, it will not be possible to establish a contract with every device or network whose services might be used in advance. Even if it would be possible, manual configuration of each connection would be a daunting task that would put away many potential users.

The problems stated above guide the discussion on adaptive security for low data rate networks, secure routing for ad hoc networks, and secure mobility in heterogeneous networks. These are discussed in the following subsections.

8.5.1 Adaptive Security for Low Data Rate Networks

The LDR networks play an important role in the following environments:

1. *Home environment*: In an apartment, up to 150 different devices could be deployed in an LDR network and managed remotely by unique controller. Star topology is best suited for this kind of piconet.

2. *Health care*: LDR networks can have a potential strong impact on health monitoring. The application that is easiest to conceive is a pacemaker implanted in a man's body, communicating heart-related information to some external device held by a doctor. In this case the network topology would be trivial, being a fixed point-to-point communication, and its implementation relatively simple. Two major challenges are: (1) devices should be very small and light, which implies extreme simplicity; and (2) big batteries are not suited for these applications.

3. *Automotive environment*: A wireless sensor network in a car could efficiently fulfill a great number of functions, which are today carried out manually or in a wired manner, negatively impacting the installation cost. It would have a star configured network, where all the sensors mounted in the car continuously transfer data to a central controller. Required data rates and maximum latency for different applications may vary dependently on the function of individual sensors.

4. *Industrial environment*: Wireless sensor networks can be used in industries for monitoring and maintenance of machines and detection of emergences, such as fire or any threshold exceeding. In this case the coverage area may be a limiting factor. A fully connected network can be achieved through a star topology and the use of repeaters, or through a hierarchical structure.

In line with the security properties and different usage scenarios, the LDR devices should support both basic and enhanced security levels. These devices can be distinguished as trusted and nontrusted devices, with the implication that a trusted device has unrestricted access to all services. Within each level, then, different choices can be made for the parameters to use. Three security levels are proposed for a nontrusted device:

1. *Low-level:* Networks providing nonprivileged services and exchanging nonsensitive data. These are generally small networks in the home environment for the remote control of domestic devices. All the networks for personal entertainment are included in this category, and they are generally made up of a reduced number of devices organized in a star topology. An efficient authentication mechanism can be simply based on an ACL. The solutions adopted for the implementation of this security level minimize the required efforts to be carried out by network's devices.

2. *Medium-level:* Networks in which some protection is needed, even if the data exchanged within the network is not necessarily sensitive. The security provided in these cases is focused on authentication and

authorization. This security level applies to networks that can be the object of active attacks, even if the data exchanged within the network is not considered highly sensitive, for example:

- Small sensor networks in the home environment;
- Large sensor networks (WINS) for the monitoring of some environmental condition over an extended territory;
- Wireless PC peripherals such as printers, cameras, calculators, mobile phones.

3. *High-level:* Networks that provide privileged access to service and/or exchange highly sensitive data. The provision of this security level implies heavy compromises with network performance, such as in health monitoring applications; for example, the smart card, which has all the personal information of an individual. Here very strong authentication and confidentiality is required. Also, established associations would be transient, which implies that the key exchange procedure should be carried out in some practical way. Finally, strong freshness would be necessary, in order to make it impossible for any reply attack.

To adapt security levels of LDR devices according to the services/applications and its requirements, a novel security manager concept has been proposed.

1. *Security level management:* The three defined levels of security allow covering the exigencies of every LDR service. First is to implement a high level of security, in order to provide the maximum protection to the transferred data, and the second is to give to security a secondary role, in order not to affect the performance of network's devices. LDR devices are generally dedicated devices. This assumption is sufficient to persuade us that it is useless to make them adaptive to diverse security needs. While these devices do not have to adapt themselves to the network, the network does need to adapt itself to the devices.

2. *The security manager:* The most efficient way in which the concept of adaptive security can be implemented is by means of a security manager, working above the link layer and controlling the security mechanism.

8.5.2 Secure Routing for Ad Hoc Networks

There are several scenarios possible for the use of ad hoc networks:

1. Ad hoc networks with infrastructure connectivity;
2. Pure ad hoc network (i.e., a network with no infrastructure connectivity);
3. Sensor type ad hoc networks with central control;

4. Machine-to-machine ad hoc network. These ad hoc networks will be made of machines that communicate to each other. An example could be vending machines that communicate with one another and send information to the warehouse. In this scenario it is assumed that the machines do not move.

5. Fixed ad hoc network. This is an ad hoc network in which the nodes are not mobile. An example could be WLAN APs or PCs. Such ad hoc networks could also be purely ad hoc or have infrastructure connectivity.

6. Military ad hoc networks.

Of the six discussed scenarios, the last one is not of interest to this chapter as the military requirements are very different. The rest of the scenarios have some similarities. The sensor type ad hoc network (scenario 3) is a network under full control, and a much more complicated form of this scenario is scenario 2. The fourth scenario could have different forms. The first is the situation in which the machines belong to one organization and they are purely ad hoc; this is similar to scenario 5, and thus to scenario 2. A second situation would be where the machines talk to an infrastructure network, thus similar to scenario 1.

There are several threats in ad hoc networks. Ad hoc networks use wireless data transmission. This makes them susceptible to passive eavesdropping, message replaying, message distortion, and active impersonation. Mobile nodes have low physical security. They can be easily compromised. This means that attacks can come also from inside the ad hoc network. Therefore, a centralized node cannot be trusted, because if this node were compromised the whole network would be useless. One problem is scalability. Ad hoc networks can have hundreds or even thousands of mobile nodes. This poses challenges to security mechanisms.

An ad hoc network is a dynamically changing multihop network that is created by the MNs when needed for their own communication purposes.

Authentication is needed in ad hoc networks. One way to deal with low physical security and availability constraints is the distribution of trust. Trust can be distributed to a collection of nodes.

Authorization is also needed to avoid a malicious host to be able to wreak havoc inside the network. It needs some sort of distributed structure. The traditional way of using ACL in one central server is not adequate.

Accounting features are quite specialized in ad hoc networks. Basically there is no network infrastructure that provides the service; nor is there the same kind of service provider concept.

When thinking about the ad hoc networks and general AAA systems, it can easily be seen that they do not fit well together. The biggest problem is related to the varying nature of the network. There are no home domains or foreign domains, because the networks come and go on demand. Also, the term "service provider" will have a different meaning than before. This does affect the AAA systems because some of the basic building blocks of their architecture are missing from the ad hoc networks. The basic problem here is that the model provided by the AAA working group is a centralized trust model. This clearly does not fit well into ad hoc networks, because the network structure is decentralized.

8.5.3 Secure Mobility in Heterogeneous Networks

Mobility management supports roaming users to enjoy their services on progress through *different* access technologies either simultaneously or one at a time. There are many distinct but complementary techniques for mobility management to achieve its performance and scalability requirements. There are two kinds of mobility: macromobility and micromobility.

The Mobile IP (MIP) protocol has been specified for wide area macromobility management. It enables a node to move freely from one point of connection on the Internet to another, without disrupting end-to-end TCP connectivity. Mobile nodes are required to securely register a care-of-address (COA) with their home agent (HA) while roaming in a foreign domain. If security mechanisms are not employed, however, the network can be compromised through remote redirection attacks by malicious nodes. In addition, mechanisms are needed that allow foreign agents (FA) in the visited domain to verify the identity of mobile nodes and authorize connectivity based on local policies or the ability to pay for network usage.

AAA functionality is needed in MIP network, because in the Internet there is quite often a need to access resources in another administrative domain. In this case the FA needs to authenticate the MN before it can give the service to MN. Typically it asks for credentials that are used to authenticate the user. FA is not able to make the decision itself, so it needs to contact the local AAA server, which may give the answer or contact another server. The purpose of the MIP AAA is to provide improvements to the basic MIP protocol:

1. Better scaling of security associations;
2. Mobility across administrative domain boundaries;
3. Dynamic assignment of HA.

The current design of MIP works well when working inside one administrative domain. But there is a need to improve the mobility across multiple administrative domains. This change will affect the current trust model of MIP. AAA tries to implement these changes and make the MIP system such that it scales well to multiple administrative domains. AAA also enables administrative domains to charge mobile nodes for the use of the network capacity. An infrastructure for AAA services to the various entities in the network can be provided.

Micromobility architectures consist of a hierarchy of nodes and base stations connected via a single point of attachment to the Internet. Cellular IP (CIP) network consists of a gateway (GW) router that connects it to the Internet as well as several CIP nodes that are responsible for routing of packets within the network. The GW represents both HA and FA functionality, and it is responsible for filtering out all signaling messages that are specific to the CIP network. Page and route update messages can be used create entries within CIP caches, which can result in changes to the routing of packets within the network. It is therefore of vital importance that each CIP node, prior to acting on any signaling information, authenticate all such messages. Unauthenticated signaling messages can be used to impersonate another node and create denial-of-service attacks. A malicious host could generate false signaling messages and trick the node's HA into adding a false COA for the node in its routing tables. This would result in packets destined for a node being routed incorrectly by the HA to an unknown destination.

References

[1]   Prasad, R., *Universal Wireless Personal Communication*, Norwood, MA: Artech House, 1998.

[2]   Prasad, R., and M. Ruggieri, *Technology Trends in Wireless Communications*, Norwood, MA: Artech House, 2003.

[3]   Ohmori, S., H. Wakana, and S. Kawase, *Mobile Satellite Communications*, Norwood, MA: Artech House, 1998.

[4]   3GPP, http://www.3gpp.org.

[5]   GSM MoU, http://www.gsmworld.com.

[6]   IEEE 802.16, available at http://ieee802.org/16/.

[7]   IEEE 802.11, available at http://grouper.ieee.org/groups/802/11/.

[8]   IEEE 802.15, available at http://grouper.ieee.org/groups/802/15/.

[9]   Stallings, W., *Cryptography and Network Security: Principles and Practice*, Second Edition. Englewood Cliffs, NJ: Prentice Hall, 1999.

[10] Stallings, W., *Network Security Essentials: Applications and Standards*, Englewood Cliffs, NJ: Prentice-Hall, 2000.

[11] Schneier, B., *Applied Cryptography: Protocols, Algorithms, and Source Code in C*, Second Edition, New York: John Wiley & Sons, 1985.

[12] http://biz.yahoo.com/rc/030903/telecoms_israel_gsm_1.html.

[13] http://www.newswireless.net/articles/031002-scam.html.

[14] http://www.checkpoint.com/*.

[15] 3GPP TS 33.102; 3G Security; Security Architecture (Release 4).

[16] IEEE P802.11i, Draft Supplement to IEEE Std 802.11, 1999 Edition, Draft Supplement to STANDARD FOR Telecommunications and Information Exchange Between Systems—LAN/MAN Specific Requirements—Part 11: Wireless Medium Access Control (MAC) and physical layer (PHY) specifications: Specification for Enhanced Security.

[17] BLUETOOTH SIG, http://www.bluetooth.com.

[18] Faeserotu, J. R., and R. Prasad, *IP/ATM Mobile Satellite Networks*, Norwood, MA: Artech House, 2001.

[19] Noubir, A., and L. von Allman, "Security Issues in Internet Protocols over Satellite," *Proceedings of IEEE VTC '99—Fall*, Amsterdam, September 19–22, 1999.

[20] http://www.gsmworld.com/using/algorithms/index.shtml.

[21] Prasad, N. R., "GSM Evolution Towards Third Generation UMTS/IMT2000," *ICPWC'99*, Jaipur, India, February 17–19, 1999.

[22] Prasad, N. R., "An Overview of General Packet Radio Services (GPRS)," *First International Symposium on Wireless Personal Multimedia Communications (WPMC'98)*, Yokosuka, Japan, November 4–6, 1998.

[23] Prasad, R., and N. R. Prasad, "GSM Evolution: Towards the EDGE," *ICPWC'99*, Jaipur, India, February 18, 1999.

[24] Ojanpera, T., and R. Prasad, *Wideband CDMA for Third Generation Mobile Communications*, Norwood, MA: Artech House, 2000.

[25] The UMTS Forum, http://www.umts-forum.org.

[26] Prasad, N. R., and K. K. Larsen, "Third Generation (3G) Networks and Standards," in S. Dixit and R. Prasad (Eds.), Norwood, MA: Artech House, 2003, pp. 45–56.

[27] Pejanovic, M., and N. R. Prasad, "Optimized Deployment of 3G/4G Mobile Systems," *World Wireless Congress 2003*, San Francisco, May 27–30, 2003.

[28] Prasad, N. R., and K. K. Larsen, "3G and Its Releases: Impact on the Existing Network," *Fourth International Symposium on Wireless Personal Multimedia Communications (WPMC'01)*, Aalborg, Denmark, September 9–12, 2001.

[29] IEEE Std 802.16-2001, "Local and Metropolitan Area Networks—Part 16: Air Interface for Fixed Broadband Wireless Access Systems."

[30] Prasad, A. R., and N. R. Prasad, *WLAN Systems: Mobility, Integration, Security and QoS*, Norwood, MA: Artech House, 2004.

[31] Prasad, N. R., and A. R. Prasad, (Eds.), *Wireless LANs Systems: Providing Wireless IP for Next Generation*, Norwood, MA: Artech House, 2002.

[32] Prasad, A. R., *Wireless LANs: Protocols, Security and Deployment*, Ph.D. Dissertation, Technical University of Delft, 2003.

[33] Prasad, A. R., H. Moelard, and J. Kruys, "Security Architecture for Wireless LANs: Corporate & Public Environment," *VTC 2000 Spring*, Tokyo, Japan, May 15–18, 2000, pp. 283–287.

[34]   Prasad, A. R., et al., "Wireless LANs Deployment in Practice," in *Wireless Network Deployments*, R. Ganesh and K. Pahlevan (Eds.), Boston, MA: Kluwer Publications, 2000.

[35]   Prasad, A. R., et al., "Wireless LANs Deployment in Practice," *International Journal on Wireless Personal Communications*, Kluwer Academic Publishers, 2001.

[36]   Prasad, N. R., "IEEE 802.11 System Design," *IEEE International Conference on Personal Wireless Communications*, Hyderabad, India, December 2000.

[37]   Prasad, A. R., et al., "Indoor Wireless LANs Deployment," *VTC 2000 Spring*, Tokyo, Japan, May 2000.

[38]   Prasad, N. R., and H. P. Schwefel, "A State-of-the-Art of WLAN and WPAN," *European Conference on Wireless Technologies (ECWT) 2003*, Munich, October 9–10, 2003.

[39]   IEEE Std. 802.1X-2001, "IEEE Standard for Local and Metropolitan Area Networks—Port-Based Network Access Control," June 14, 2001.

[40]   IEEE P802.15 Wireless Personal Area Networks, IEEE P802-15_TG4 NTRU Security Architecture Proposal, June 12, 2002.

Part 3
The Unpredictable Future

9

WPAN as a Part of 4G

9.1 Introduction

In the view of the authors, 4G can be defined by the following equation:

$$B3G + \textit{Pers} \overset{\Delta}{=} 4G \tag{9.1}$$

where B3G stands for beyond third generation, which is defined as the integration of existing systems to interwork with each other and with the new interface. *Pers* stands for personalization, and this topic is under research in the European Commission 6th Framework Program within an Integrated Project, My Personal Adaptive Global Net (MAGNET: http://www.ist-magnet.org).

Current telecommunications and computer networks are on the verge of providing mobile multimedia connectivity, where nomadic users would have ubiquitous access to remote information storages and computing services. As an evolutionary step toward B3G, mobility in heterogeneous IP networks with both UMTS and IEEE 802.11/802.16 WLAN systems is seen as one of the central issues in the future B3G construction

In the future, mobile access to the Internet will be a collection of different wireless services, often with overlapping areas of coverage. A single technology or service cannot provide ubiquitous coverage, and it will be necessary for a mobile terminal to employ various points of attachment to maintain connectivity to the network at all times. The most attractive solution for such consideration is to utilize high bandwidth data networks such as IEEE 802.11WLAN, Bluetooth/802.15 WPANs, and emerging 802.16/20 systems whenever they are available, and switch to an overlay public network such as UMTS/HSDPA with lower bandwidth when there is no WLAN coverage. Think of a scenario where users may wish to be connected to WLAN for low cost and high bandwidth in the home, airport, hotel, or shopping mall, but also want to connect to cellular technologies (e.g., GPRS or UMTS) from the same terminal. In particular, the users in this scenario require support for vertical handover (handover between heterogeneous technologies) between WLAN and UMTS.

Our vision, shared by the MAGNET project is that personal networks will support the users' professional and private activities, without being obtrusive and while safeguarding their privacy and security. A PN can operate on top of any number of networks that exist for subscriber services or are composed in an ad hoc manner for this particular purpose. These networks are dynamic and diverse in composition, configuration, and connectivity depending on time, place, preference, and context, as well as resources available and required, and they function in cooperation with all the needed and preferred partners. As such, our belief is that cellular networks as known today will be an important part of these PNs, but people will consider the PNs in the same way they consider today's mobile phone networks: an indispensable tool that we call 4G.

Sections 9.2 and 9.3 describe the general architecture and challenges of personal networking, inspired by the MAGNET project. Sections 9.4 and 9.5 give examples of how interconnection and handover between WLAN and UMTS technologies can be implemented. Section 9.6 recaps the most salient features of what 4G will be.

9.2 Personal Networks: Definition

In a personal distributed environment, users interact with various companions, embedded or invisible computers not only in their close vicinity but potentially anywhere. We call these systems personal networks after [3, 4]. They constitute a category of distributed systems with very specific characteristics [1–14].

PNs comprise potentially "all of a person's devices capable of network connection whether in his or her wireless vicinity, at home or in the office." The work towards realizing this vision has resulted in major extensions of the present PAN and ambient networking (AN) paradigms. PNs are configured in an ad hoc fashion, as the opportunity and the demand arise to support personal applications. PNs consist of communicating clusters of personal and foreign devices, possibly shared with others, and connected through various suitable communications means. At the heart of a PN is a core PAN, which is physically associated with the owner of the PN, as illustrated in Figure 9.1 by the area covered by dotted line. Unlike PANs, with a limited geographic coverage, PNs have an unrestricted geographical span and may incorporate devices into the personal environment regardless of their geographic location. In order to extend their reach, they need the services of infrastructure-based, and possibly also ad hoc,

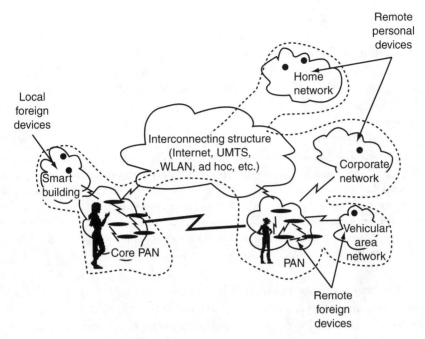

Figure 9.1 Illustration of the PN concept.

networks. A PN is the essential communication element that will help make the concept of pervasive computing a reality.

The main components of a PN are as follows:

- A *core PAN* consisting of personal devices in the close physical vicinity of a user, including devices moving around with him or her. The core PAN is an essential component of the PN.

- *Local foreign devices* or *clusters* thereof, which are owned by other parties and could either be reserved solely for the PN owner or be shared with others. They are linked to the core PAN via communication infrastructure.

- *Remote personal devices*, which are grouped into cooperating clusters and which are linked to the core PAN via communication infrastructure.

- *Remote foreign devices* or *clusters* thereof, which are linked via communication infrastructure, and again can be shared with many users or be reserved for the PN owner.

- *Communication infrastructures*, in principle WANs making use of some sort of infrastructure-based resources (CN), which can be public (e.g., cellular, Internet) or private (e.g., leased lines), licensed or unlicensed (e.g., WLAN).

The PN has to support resource-efficient, robust, ubiquitous service provisioning in a secure, heterogeneous networking environment for nomadic users. Of paramount importance is the requirement that a PN will support its owner in all her private and business activities, without being obtrusive and while safeguarding the security and privacy of the users and their data.

A simple example is a PN-based remote babysitting application (Figure 9.2) [4]:

> Consider the case of a mother visiting a friend's house while her child is asleep at home (supervised by a person not specialized in childcare). She might want to remotely watch and observe the child. She does this by using a PN consisting of some personal devices, e.g., a UMTS and Bluetooth-enabled PDA and a headset she carries with her, and, a remote pair of eyes and ears in the child's bedroom at home. The latter consist of a digital video camera, a microphone and a UMTS phone, forming a cluster of cooperating devices.

Alternatively, since the friend's living room is equipped with a TFT wall display including speakers, hooked up to a home network and accessible to authorized guests via a Bluetooth, the mother may want to use these instead to observe the child.

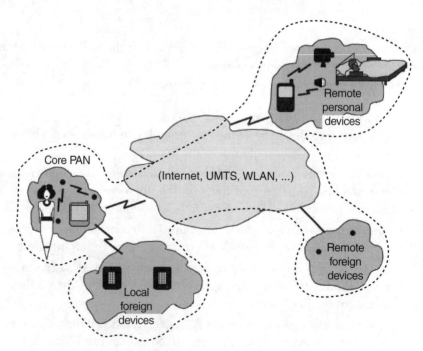

Figure 9.2 Remote babysitting application [4].

9.3 Personal Networks: The Challenges

The development of PNs requires the solution of a number of technological issues related to networking, coexistence and interworking between a multitude of different network interconnection schemes, wireless technology for PNs, security, and privacy. PNs also have to be developed with user and socioeconomic issues in mind, as described in Chapter 5, to ensure that the concept is validated from a business and user perspective.

The most challenging research topics in the development of PNs are as follows:

- The development of user-centric business and usage model concepts for PNs in multinetwork, multidevice, and multiuser environments, leading to user requirements definition;
- The design and development of an architecture and protocols for building PNs based on heterogeneous networks, and optimized from a user's perspective;
- The development and validation of efficient, flexible, and scalable air interface(s) for the PAN components of a PN;
- The development of integrated mechanisms for a secure PN in order to ensure security, trust, and privacy of the user's data, including in all communication, networking, and application aspects;
- Overall integration by cross-layer optimization of lower and upper layers, including all low power aspects and support of end-to-end QoS.

Even if the PN concept above fits with the visions produced by different groups and from different perspectives (the IST Advisory Group [5], WWRF [6], KTH [7], the U.K. government's Foresight Initiative [15], the U.K. Mobile Virtual Centre of Excellence [16], EURESCOM [17], and the Association of Computing Machinery [1]), there are no published results on PNs as envisioned by the MAGNET project. However, there are a number of approaches and research initiatives that are of relevance to PNs, and we provide a brief survey of these below.

9.3.1 User Requirements

The dominant approach to user requirements is that services and applications be shaped by the combined influence of terminals and networks developed according to current technological possibilities (i.e., user requirements are not taken into account during initial conception). Service development then involves the

PAN and a combination of networks: PSTN, cellular networks, digital broadcasting networks, as well as Bluetooth, WLAN, and the Internet.

MAGNET's approach to user requirements is different. The methodology to describe and develop an understanding of implementation of an efficient PN-solution in a heterogeneous, multimodal environment involves *technology, user needs*, and *economics*. A key element of user needs is *perceived quality of service (quality of experience)* associated with given private and/or business activities and its relation to the underlying technologies. The introduction of PN services along with the associated technologies will constitute a major paradigm shift. There are currently no business models or scenarios in place for PNs; however, an enhanced understanding and knowledge of possible business model solutions as well as market and socioeconomic aspects are necessary in order to achieve the full benefits of a heterogeneous communication model as proposed in the PN concept.

9.3.2 Networking

A number of system aspects are important when addressing PN networking issues: middleware for mobile distributed systems, resource and context discovery, addressing and routing, self-organization, mobility management of subnetworks, and service discovery and provisioning in heterogeneous environments.

Middleware for mobile distributed systems. Middleware architecture involving a PN-like architecture has been developed in the MOPED project [4]. This architecture solves many problems by relying on a dedicated infrastructure-based proxy, like a home agent in Mobile IP. As indicated in Section 9.4, we believe that PNs avoid the dependence on infrastructure.

PNs also rely on an architecture for peer-to-peer computing, like Sun's JXTA,[1] that is based on an overlay network model and protocols for service discovery, security, routing, and so forth, taking into account the restrictions and characteristics of future mobile pervasive computing devices.

Resource, context, and service discovery. A resource description system (e.g. INS developed at MIT) is an essential element of resource discovery. IETF Service Location Protocol (SLP), Sun's Jini service directory,[2] the Simple Service

---

1. JXTA is short for juxtapose, as in side-by-side. JXTA technology is a set of open protocols that allows any connected device on the network, ranging from cell phones and wireless PDAs to PCs and servers, to communicate and collaborate in a P2P manner.

2. Jini is a set of Java API's defining a directory service listing references to electronic services. In Jini terminology, a directory service is a lookup (or registry). References are generally proxies for service providers registered with the lookup.

Discovery Protocol (SSDP), universal plug-and-play, and Berkeley's service discovery service are existing protocols, but they are not suitable for PNs because they do not support the highly dynamic context to be expected in PNs and preclude ad hoc operation since they involve infrastructure-based servers. IBM's DEAPspace[3] project has developed a resource discovery mechanism, which takes these considerations into account. Context-aware applications exploit information about, for example, the geographical location, the time of day, the available equipment, the interaction history, and the presence of other people, in order to provide the user with the service that is best suited to user's circumstances.

Self-organization. IEEE 802.11 provides elementary link level self-organization. For Bluetooth networks, which are likely to be an important link technology for the device clusters constituting a PN, the self-configuration of so-called scatternets (see Chapter 3) consisting of multiple piconets is still a research topic. For the network level the problem has been studied extensively. Here, however, the problem has a totally different dimension, and scalability becomes an issue.

Addressing and routing. Many routing strategies have been devised and analyzed [6]. For example, long distance geographic routing is a technique that relieves the nodes from keeping volatile network state information about distant nodes and links.

Mobility management of subnetworks. New solutions are needed when dealing with the mobility of terminal devices and subnets. Worth mentioning in this context are the activities on mobile networks within the Mobile IP Working Group [5] of the IETF and the work on extensions of Mobile IP for mobile ad hoc networks interconnection [7]. The PN also relies on interworking solutions between (and global roaming solutions to) all access technologies, with horizontal and vertical hand-over and seamless services provision.

Ad hoc networks have received a lot of attention in recent years (see Chapter 7) and are an important part of PNs. Still, the extension of MANET concepts to heterogeneous networks is ongoing and central to the development of PNs.

Building an *automated and context-aware* solution for the PNs is a very difficult research task (and is causing interest among a large number of researchers in the pervasive computing community), but it has a strong industrial potential, since it would offer the possibility of building a whole new class of applications, services, and devices for the mass market.

---

3. The aim of DEAPspace is to connect nomadic and pervasive devices in (transient) ad hoc networks to allow available resources to be utilized and coordinated towards more useful applications than any one of the component devices would be capable of supporting. DEAPspace addresses peer-to-peer networking of pervasive devices instead of client-server networking.

9.3.3   Adaptive and Scalable Air Interfaces for PANs

The PN concept involves the development and provision of a highly adaptive and spectrally efficient PAN air interface. PNs will utilize and enhance existing air interfaces and develop novel air interface technologies; they will also provide an interworking structure in the form of a Universal Convergence Layer to enable an adaptive and spectrally efficient solution across legacy PAN technologies.

The IST project MATRICE and NTT DoCoMo have investigated the feasibility of multicarrier (MC) techniques as a potential candidate for B3G. MC-techniques can accommodate advanced signal processing schemes that enhance user throughput. Combinations of advanced signal processing techniques such as space-time coding, beam forming, array processing, multiuser detection, interference cancellation, and synchronization algorithms are investigated in MATRICE, along with adaptive link layer techniques and layer interface issues. Encouraging results have motivated interest into investigating the applicability of MC-techniques for PAN networks, as a possible candidate for multiple access for terminal-to-gateway connectivity. For the PAN transceiver, additional schemes like adaptive bit and power loading strategies and adaptive coding in combination with MIMO capabilities are mandatory for the flexibility and bandwidth efficiency required in future PANs, particularly for operation in the ISM bands. The complexity of MC air interfaces for PANs and their feasibility especially with respect to processing power and power consumptions are critical research topics.

UWB techniques are well suited for short-range communications, and are addressed in several European projects as well as in the IEEE 802.15 standardization groups. Although not envisaged direction in the WPAN context, MIMO techniques will also be a part of the PNs (e.g., like that foreseen in IEEE 802.11n), since in this PN concept, the border between WPAN and WLAN is quite thin. Channel characterization in these environments is still a research topic.

Interference of UWB systems or coexistence of any air interface is a major issue and is currently being studied for fixed access mobile architectures. Due to the innovative ad hoc nature of the PN concept, study of the interference issues is required to ensure proper coordination as well as to minimize the impact on other systems operating in the same frequency bands.

9.3.4   Security

Current security solutions for wireless technologies—such as the one from 3GPP for GSM/UMTS based on (U)SIM algorithms, the IEEE 802.11i standard for WLAN security, or the Bluetooth security recommendations—are all tailored for securing the traffic exchanged between user devices and access points.

Applications running there need to secure their traffic using application-layer schemes such as TLS or IPsec. While combining both transport and application layers would result in end-to-end secure communication, this approach was optimized for the Internet with a fixed infrastructure (i.e., very low BER, fixed links, which can either be available or unavailable, and in the former case (temporarily) congested or not, and powerful end devices capable of decrypting the secure message). To achieve the same level of security in ad hoc networks, approaches based on cooperative authorization and distributed key management are being discussed. However, these solutions usually incur considerable overhead in terms of signaling (and thus bandwidth usage) and processing needs. In the case of mobile networks, the first standardization efforts are being carried out in the MANET and NEMO groups at IETF. Other IST and FP6 projects have contributed and still contribute to this area. However, a general solution that is both adaptable to the network conditions and end system capabilities as well as enabling interdomain AAA negotiation is largely still missing and is one of the tasks of MAGNET, in liaison with other FP6 projects.

9.4   A Sample Interworking System Architecture

In this section we show a solution for interconnecting a IEEE 802.11 WLAN to a UMTS core network (CN) as an example of an interworking system architecture. The communication world is converging towards all-IP solutions, as well as a domination of Ethernet-based solutions at the lower layers, so this case is a good example of what a generic interworking architecture could be in the frame of PN.

By connecting the IEEE 802.11 WLAN to a UMTS CN as a complementary radio access network, a second form of mobile packet data services is provided by this heterogeneous IP-based system. Figure 9.3 represents five possible interconnection points between WLAN and UMTS. These interconnection architectures involve minimum changes to the existing standards and technologies, and especially for the MAC and PHY layers, to ensure that existing standards and networks continue to function as before. The first two interconnections in Figure 9.3 will always have interaction between WLAN AP and the packet switched (PS) part of the UMTS CN [18]. This means that the gateway to the IEEE 802.11 WLAN network is attached to the PS domain. This interconnection is possible through the 3G serving GPRS supporting node (3G-SGSN) entity and gateway GPRS supporting node (GGSN) entity, which are the elements of the UMTS PS CN. In both cases the WLAN network appears to be a UMTS cell or routing area, respectively. The UMTS network will be a master network and the IEEE 802.11 WLAN network will be a slave network. This means that the mobility management and security will be handled by

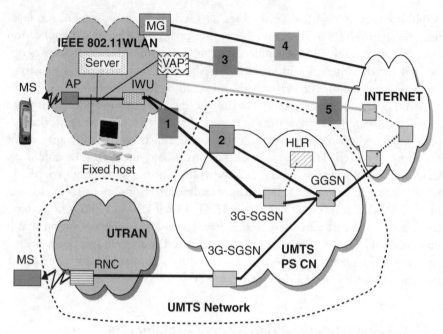

Figure 9.3 Interconnection architectures between IEEE 802.11 WLAN and UMTS. (IWU: interworking unit; MG: mobility gateway.)

UMTS network, and the WLAN network will be seen as one of its own cells or routing areas. This may require dual mode PCMCIA cards to access two different physical layers. In addition, all traffic will first reach the UMTS 3G-SGSN or 3G-GGSN before reaching its final destinations even if the final destination were to be in the WLAN home network. Hence, solutions 1 and 2 are not really viable.

In the third interconnection the virtual access point (VAP) reverses the roles played by the UMTS and WLAN in the first two architectures. This is called a tight coupling because there is always interaction between both networks. Here, the IEEE 802.11 WLAN is the master network and the UMTS is the slave network. Mobility management is according to the WLAN, and IAPP is the protocol that is specified for this management. In this solution, the protocol stack becomes quite large; in particular, it uses systematically two layered TCP/IP layers, hence incurring a lot of overhead and bandwidth waste. This solution does not appear as viable.

In the fourth interconnection architecture a mobility gateway/mobile proxy (MG) is employed between the UMTS and IEEE 802.11 WLAN networks, which both act as peer-to-peer networks. The MG is a proxy that is implemented on either the UMTS or the WLAN sides, and it will handle the mobility and routing. In this case, even if most of the problems of the first three solutions are solved, the MG introduces a significant additional latency. Furthermore, the network architecture (i.e., placement of proxies, number) is quite difficult to design.

The fifth interconnection architecture is based on the Mobile IP protocol. This is called "no-coupling" and both networks are peers. Mobile IP handles the mobility management, with the help home agent(s) (HA) and Foreign agent(s) (FA), which advertsie mobility to the mobile nodes. The mobile nodes receive a foreign address, also called core of address (COA), which is linked to their new location [19]. This appears to be the best solution and we give some details about it below.

The interconnection architecture related with the Mobile IP is presented in Figure 9.4. Mobile IP is used to restructure connections when a mobile station roams from one data network to another. Outside of its home network, the mobile station is identified by COA associated with its point of attachment, and a colocated FA that manages deencapsulation and delivery of packets [1, 2].

The MS registers its COA with an HA. The HA resides in the home network of the MS and is responsible for intercepting datagrams addressed to the MS's home address as well as encapsulating them to the associated COA. The datagrams to an MS are always routed through the HA. Datagrams from the MS are relayed along an optimal path by the Internet routing system, though it is possible to employ reverse tunneling through the HA.

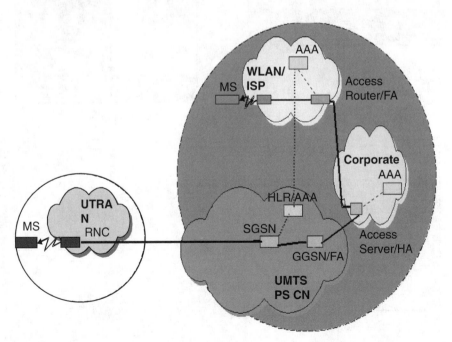

Figure 9.4 Interconnection architecture between WLAN and UMTS based on Mobile IP.

The required dual mode MS protocol stack is given in Figure 9.5. It is clear that both networks are peer networks and that the functionality of the HA/FA exists at the IP layer [2].

The advantage of this interconnection is that it is based upon the Mobile IP protocol, with a possible extension to IPv6. The same IP address is used, which solves the multiple address problems. In order to solve the packet duplications due to the lifetime of the routers, some conventions on both IEEE 802.11 WLAN and UMTS networks are needed. The databases of both networks may need to communicate to overcome packet duplication.

The main disadvantage of Mobile IPv4 is the triangle routing. This could be overcome with Mobile IP with optimized routing and is important for real-time applications like video/audio transmission.

Industrial alliances like Unlicensed Mobile Access (UMA) and Seamless Converged Communication Across Networks (SCCAN) are proposing solutions for seamless handover between WLANs and cellular technologies.

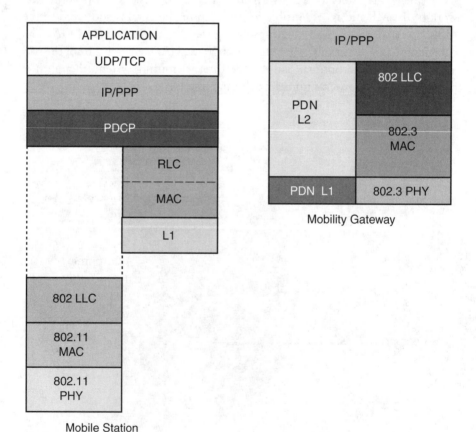

Figure 9.5 Protocol stacks corresponding with the MS and the MG.

9.5 A Handover Example

Like in the previous section, we introduce a handover example between 802.11 WLAN and UMTS. In the frame of PNs, a generic handover mechanism will have to be designed and applied to all standards that are at stake. This example also shows that large standardization efforts will have to take place to enable the use of large heterogeneous networks.

The motivation intertechnology (vertical handover) for the hybrid mobile data networks arises from the fact that no one technology or service can provide ubiquitous coverage, and it will be necessary for a mobile terminal to employ various points of attachment to maintain connectivity to the network at all times. There is a clear difference between the two types of handover, namely, horizontal and vertical handover. Horizontal handover refers to handover between node Bs or APs that are using the same kind of network interface. Vertical handover refers to handover between a node B and an AP, or vice versa, that are employing different wireless technologies. Vertical handover can be categorized as:

- Upward vertical handover, which occurs from IEEE 802.11 WLAN AP with small coverage to an UMTS node B with wider coverage;
- Downward vertical handover, which occurs in the reverse direction.

A downward vertical handover has to take place when coverage of a service with a smaller coverage than in WLAN service becomes available when the user still has a connection to the service with the UMTS coverage. An upward vertical handover takes place when an MS moves out of the IEEE 802.11 WLAN coverage to UMTS services when they becomes available, but still the user has a connection to the IEEE 802.11 WLAN coverage. In the case of the vertical handover, the mobile station/host decides that the current network is not reachable and hands over to the higher overlay UMTS network, when several beacons from the serving WLAN service are not available. It instructs the WLAN to stop forwarding packets and routes this request via the Mobile IP registration procedure through the UMTS core network. When it is connected to the UMTS network, the mobile station listens to the lower layer WLAN AP and if several beacons are received successfully, it will switch to the IEEE 802.11 WLAN network via the Mobile IP registration process. Thus the vertical handover decisions are made on the basis of the presence or absence of beacon packets.

Handover is the mechanism by which an ongoing connection between the MS and the connecting host (CH; can be mobile or not) is transferred from one point of access to another point while maintaining the connectivity. When an MS moves away from an AP or from a node B, the signal level degrades and there is a need to switch communications to another point of attachment that gives access to the existing IEEE 802.11 WLAN network or UMTS network. The handover mechanism in an overlay UMTS and underlay WLAN network could be performed such that the users attached to the UMTS just occasionally check for the availability of the underlay WLAN network. A good handover algorithm is needed to make the decision when to make handover in order to avoid unnecessary handover (i.e., ping-pong effect). The handover procedure and the mechanism from WLAN to UMTS (and vise versa) are based on the received signal strength (RSS) metrics. This means that the handover initiation or the handover triggering is sensitive to these signals [7]. Figure 9.6 shows the handover procedure from one network to another.

An MS moving from the WLAN network coverage may suddenly experience severe degradation of service and will have to perform handover very quickly to maintain the higher layer connection. The following stages occur when an MS moves away from the coverage of WLAN within the UMTS coverage (Figure 9.7).

1. The signals received from the AP in the WLAN network are initially strong and the MS is connected to the WLAN network, which is also the home network of the MS. The HA also resides in this network.

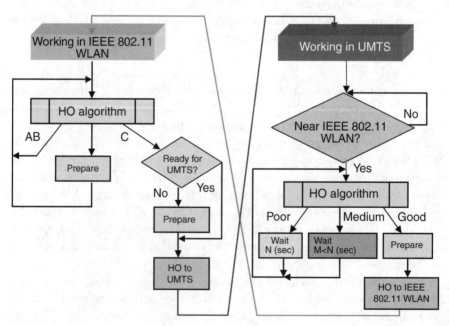

Figure 9.6 Handover procedure between WLAN and UMTS.

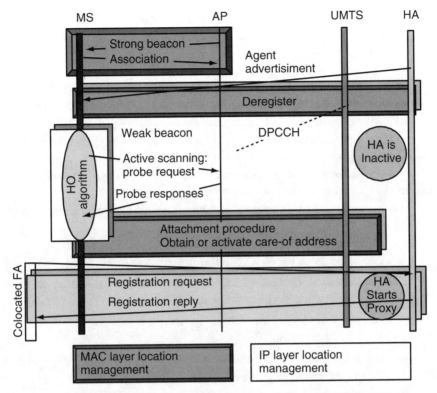

Figure 9.7 Messages and signaling of the handover procedure from IEEE 802.11 WLAN to UMTS.

2. The signals from the AP become weaker when the mobile moves away. The MS scans the air for other APs. If no AP is available, or if the signal strength from the available AP is not strong enough, the handover algorithm uses this information along with other possible information to make a decision on handing over to higher overlay UMTS network. Connection procedures are initiated to active the UMTS PCMCIA card.

3. The handover algorithm in the MS decides to dissociate from the WLAN and associate with the UMTS network.

4. The FA is activated used by the MS dual PCMCIA card and the Mobile IP protocol, and the MS gets a COA because it is visiting the UMTS network as a foreign network.

5. The HA in the WLAN is informed about the new IP address through a Mobile IP registration procedure and it does proxy Address Resolution Protocol (ARP) and intercepts the datagram. The HA encapsulated the

datagrams and tunnels any packets arriving for the MS to the FA of the UMTS networks. At the end of the delivery, the MS will be deencapsulated and get the datagrams.

In this case, the handover algorithm determines that there is no local coverage available via WLAN, and so handover must be performed to the UMTS network, assuming that a UMTS service is always available to the MS.

Once the MS is attached to the UMTS, it constantly monitors the air at repeated intervals to see whether or not a high data rate WLAN service is available. As soon as such a service becomes available, the handover algorithm should initiate an association procedure to the newly discovered AP.

The procedure for this reverse handover from UMTS to WLAN network is as follows:

1. The signal from the WLAN AP is initially not detected.

2. The MS then detects a beacon, which indicates that the underlay WLAN network has become available.

3. The handover algorithm decides to make a handover from UMTS to the WLAN network.

4. The FA in the UMTS network is deactivated and updated by Mobile IP, and the home IP address is used.

5. The HA in the WLAN network is instructed by the MS to no longer do a proxy ARP on its behalf through the Mobile IP protocol.

9.6 Conclusions

PAN is an emerging paradigm that has attracted an increasing amount of activity not only in academia, but also in industry. The prevalent approach to PANs concentrates on short-range communications with limited communications capabilities (e.g., Bluetooth, which is a master-slave system with scatternet extensions). WLAN developments have in the meantime made it possible to build more sophisticated services based on short-range communications.

PN extends the state-of-the-art PANs that interact in a wireless fashion within the user's *personal bubble*. PN needs optimized short-range wireless protocols and, instead of using large scale ad hoc routing, relies on interoperability and interworking of routing mechanisms between core networks and PNs (including addressing schemes and service discovery). Another novelty is the objective of having PN networks that are seamlessly interoperable not only with any access technology, but also, through any type of core network, be able to connect to other PNs. This means that PN-to-PN communication links need to be established that are secured through, for example, automatic VPN tunneling in the IPv6 context. Finally, some active networking and agent brokerage structure could be used at the network edges to provide better flexibility and optimization.

4G, a convergence of PAN/WLAN/cellular and fixed network, will offer a personalized and pervasive network to the user. Although the challenges involved with moving toward this vision of 4G are still significant, the huge amount of research and development occurring around the world in all of the areas pinpointed in this chapter, as well as the projects working on the convergence of these technologies, are moving us towards the advent of a really exciting and disruptive concept of fourth generation mobile networks.

References

[1]   "The Next 1000 Years," special issue of *Communications of the ACM*, Vol. 44, No. 3, March 2001.

[2]   Kravets, R., C. Carter, and L. Magalhaes, "A Co-Operative Approach to User Mobility," *Computer Communication Review*, Vol. 31, No. 5, October 2001, pp. 57–69.

[3]   Niemegeers, I. G., and S. M. Heemstra de Groot, "From Personal Area Networks to Personal Networks: A User Oriented Approach," *Personal Wireless Communications*, Kluwer Journal, May 2002.

[4]   Niemegeers, I. G., and S. M. Heemstra de Groot, "Personal Networks: Ad Hoc Distributed Personal Environments," invited paper at *Med-HocNet, IFIP Conference on Ad-Hoc Networks*, September 2002.

[5]   Ducatel, K., et al., (Eds.), "Scenarios for Ambient Intelligence in 2010," IST Advisory Group (ISTAG), European Commission, Brussels, http://www.cordis.lu/ist/istag.htm, 2001.

[6]   Mohr, W., et al., (Eds.), "The Book of Visions 2000," Version 1.0, Wireless Strategic Initiative, November 2000, available at http://www.wireless-world-research.org/general_info/Bookofvisions/BookofVisions2000.pdf.

[7]   Zander, J., et al., "Telecom Scenario's in 2010," internal report, KTH, Sweden, 1999.

[8]   IETF MANET Working Group, http://www.ietf.org/html.charters/manet-charter.html.

[9]   Perkins, C. E., *Ad Hoc Networking*, Reading, MA: Addison Wesley, 2001.

[10]   Jönsson, U., et al., "MIPMANET—Mobile IP for Mobile Ad Hoc Networks," *Proceedings of the IEEE/ACM Workshop on Mobile and Ad Hoc Networking and Computing*, Boston, MA, August 2000.

[11]   Lilleberg, J., and R. Prasad, "Research Challenges for 3G and Paving the Way for Emerging New Generalization," *Wireless Personal Communications*, No. 17, 2001, pp. 355–362.

[12]   Prasad, R., and M. Ruggieri, *Technology Trends in Wireless Communications*, Norwood, MA: Artech House Publishers, 2003.

[13]   Hara, S., and R. Prasad, *Multicarrier Techniques for 4G Mobile Communications*, Norwood, MA: Artech House, 2003.

[14]   Prasad, R., and L. Munoz, *WLANs and WPANs Towards 4G Wireless*, Norwood, MA: Artech House, 2003.

[15]   UK Foresight, http://www.foresight.gov.uk/.

[16]   Core 3 Programme, Mobile Virtual Centre of Excellence, http://www.mobilevce.com/.

[17]   "Systems Beyond 3G—Operators Vision," *Seventh WWRF Meeting*, Einhoven, The Netherlands, December 2002.

[18]   Prasad, N. R., "An Overview of General Packet Radio Services (GPRS)," *Wireless Personal Multimedia Communications (WPMC'98)*, Yokosuka, Japan, November 4–6, 1998.

[19]   IP Mobility Support for IPv4, available at http://www.ietf.org/rfc/rfc3220.txt.

10

PN Applications

10.1 Introduction

A personal network can be considered as an ordinary personal area network, but without geographical limitations. In a PAN all the devices are within a certain distance from each other (up to 10m), see Figure 10.1. This gives certain suggestions of connection technologies used in the PAN (Bluetooth, IR, etc.). In a PN devices can be separated by hundreds of kilometres and still belong to the same virtual PN. It means that in a PN, connection technologies are not only short range like Bluetooth or WLAN, but also medium range and national or continental range, like GPRS or UMTS. This calls for more service requirements for the PN as compared to the PAN. These are especially different when we talk about technical requirements, because some issues in a PN are much more complex to achieve than in a PAN. Seamless service, mobility, single sign-on, context discovery, self-organization, roaming, handover, context transfer, and session continuity are the requirements that will be the main technological challenges for PN. If these issues could be realized with the same quality as in a PAN, then there would not be much difference from the service point of view between the PN and PAN concepts. The physical network structure would be different but the logical structure would remain the same.

A key feature of the PN is its emphasis on the trust relationship between the user and the devices. In a PN environment, the private PAN (P-PAN) contains the most important devices or device group of the user. The P-PAN devices

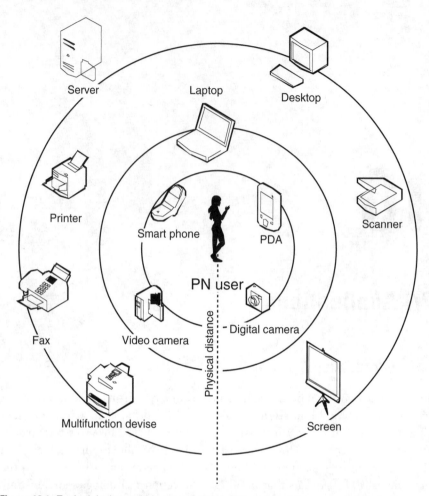

Figure 10.1 Typical devices within a user's PAN. Only a part of these may belong to the user's P-PAN

are within the user's PAN, but the PAN may also contain other devices. The user has 100% trust in his/her P-PAN and normally stores personal information (preferences) in it, which might be very sensitive. How can this trust relationship be guaranteed at any time, and what should users do when they enter new environments and need to add new devices to their P-PAN? The mobility of the PN users and continuous maintenance and updates of trust relationships represent the major challenges for the PN service architecture.

The P-PAN, as defined in the MAGNET [1] project, does not have any strict physical coverage or any limitation on the number of devices within it. It is a kind of logical concept that represents the level of trust from the user to the device. This means that in the PN environment the P-PAN of a specific user may shrink or spread its physical coverage, according to the trust relationship from the user to the devices within or outside the user's vicinity. For instance, in the mobile gaming case a teenage PN user, say Ruchika, might carry around a PDA that belongs to Ruchika's daily P-PAN. When Ruchika enters a game store and decides to try a new game, she might have to leave her PDA in the locker outside the game area due to safety or other consideration. Still she could synchronize her PDA with the intelligent device offered by the store and authorize the intelligent device as her temporary P-PAN.

The P-PAN is inherently dynamic, and the number of devices may change. But some devices are likely to be so important that they are always a part of the P-PAN. The importance and the distribution of the devices will be determined by the applications of the PN.

There are several applications of PN, the generic list of service themes includes:

1. Shopping;
2. Education;
3. Travel;
4. Community;
5. Collaborative work;
6. Surveillance;
7. Emergency;
8. Health care;
9. Society–citizen services;
10. Entertainment;
11. Transportation.

Table 10.1 illustrates the cross section of the possible use cases against the corresponding service themes [2–8]. This chapter is organized as follows. Section 10.2 presents the health care theme, describing a diabetes case and an emergency case. The shopping theme is introduced in Section 10.3, which describes smart shopping. Section 10.4 discusses the education theme by introducing the student case. The entertainment theme considers mobile gaming and is presented in Section 10.5. Section 10.6 considers distributed work as a theme and presents use cases of nomadic users, namely, washing machine repair service, virtual home truck, journalism, and remote printing service. This chapter concludes with Section 10.7.

Table 10.1
Theme Coverage of All Cases

Diabetes Case	Smart Shopping	Washing Machine	Student	Mobile Gaming	Distributed Work	Digital Living	Whole Person	Virtual Home Truck	Remote Printing	Emergency Case
Transportation	X					X				
Entertainment		X		X	X		X	X	X	
Society/citizen	X						X	X		
Health care	X									
Emergency	X									X
Surveillance	X		X	X			X	X	X	X
Collaborative work			X	X	X	X		X	X	
Community	X	X		X	X	X	X	X	X	X
Travelling	X					X			X	
Education		X		X						
Shopping								X		

10.2   Health Care

Health care is certainly a scenario that would benifit tremendously from the enhanced usage of PN. A patient's health can be monitered by doctors remotely, implicitly reducing the financial and nonfinancial operational costs of hospitals. Doctors can devote more time for the patients who require urgent health care. This section will address a diabetic and emergency personal network driven scenario [1].

10.2.1   Diabetic Case

This section describes how diabetic patients can lead a "normal life" by not having to visit the doctor frequently. The PN of the patient would enable the doctor and other concerned parties to monitor the health of the patient. The project Diabetes Advisory System—Personal Networked (DiasNet-PN) is looking into these issues.

The aim of DiasNet-PN is to develop an information and communication technology (ICT)-based user-centric service that helps involve insulin dependent (type-1) diabetics in disease self-management. DiasNet-PN is a public health care service prototype functioning within the framework of MAGNET. Figure 10.2(a) illustrates how DiasNet-PN generally is envisioned in a MAGNET context, and Figure 10.2(b) shows a close up of the patient's P-PAN.

DiasNet-PN enables patients to enter retrospective data on carbohydrate intake, insulin injections, and blood glucose readings. The data is entered into a wearable P-PAN device, which in the future is built discretely into a wrist-watch or a standard mobile phone; for now, the device will be referred to as the intelligent monitoring device (IM-device). The IM-device communicates data via manual user input and automatically via the wearable intelligent blood glucose meter (IBG-meter) and via the intelligent insulin pen (II-pen) [see Figure 10.2(b)]. The IBG-meter continuously measures the glucose concentration and wirelessly transmits the readings to the II-pen and the IM-device.

An appropriate insulin dosage is calculated by the IM-device and the results are forwarded to the II-pen so the patient can administer the correct insulin dosage. If the patient decides to administer the advised insulin dosage, the IM-device logs the dosage; if the patient decides to administer a dosage different from the advised one, the IM-device logs both the advised and the administered dosage. Manual control overrules technical automatisms at all times, and robustness, privacy, and reliability are of utmost importance.

If the patient decides not to wear the IM-device for a while (e.g. when sleeping or participating in sports), the vibrating receiver device (VR-device) can be used as a substitute. The VR-device is a physically robust and small device able to notify the diabetic of special events (e.g. low blood glucose

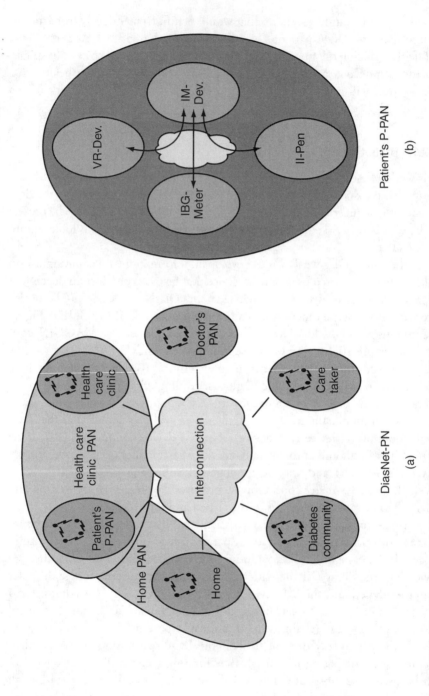

Figure 10.2 (a) Envisioned DiasNet-PN in a MAGNET context and (b) Patient's P-PAN close up.

concentration). Vibrations, sound, light, or a combination will give notification of different events.

Based on the data provided to DiasNet-PN via the IM-device, a graphical and tabular overview of the data is returned. If information about expected future carbohydrate intake is provided manually by the user, the IM-device will also estimate the future blood glucose profile and appropriate insulin dosages. Hereby, the patient has the possibility to experiment with data and learn how to optimize future insulin dosages according to carbohydrate intake.

If the patient encounters a specific problem, or has a question that cannot be handled directly by DiasNet-PN, it is possible via the IM-device to contact the associated professional diabetes team at the health care clinic. A professional diabetes team typically consists of a medical doctor, a nurse, a dietician, and a secretary.

10.2.2 Emergency Case

The emergency health care case (EC) involves actions and accompanying measures needed for starting up suitable treatments of sudden, critical, medical or physiological cases that demand immediate action and treatment by health care personnel [2–4].

Scenario Description

The EC covers cases that include patients from traffic accidents, accidents in the home or at the workplace, people suffering from heart attacks or shock, and victims of natural or man-made catastrophes, such as fires and earthquakes. Fast treatment can be supported by access to a large number of information resources (e.g., the person's electronic patient journal transmitted to the ambulance on the way from the incident location and the hospital) and by scanning the patient during transportation using available network connection [5].

The main focus of EC is to conduct initial treatment of the patient, enable the fastest possible transport of the patient from the location of incident to the hospital, and to have as much immediate communication between the transport vehicles (ambulance or helicopter) as possible. The people involved in this scenario include:

- The patient(s);
- Persons in close vicinity of the patient;
- Paramedics (and persons) called to perform initial treatment and provide transportation to a hospital;
- Doctors and other health care personnel at the receiving hospital, specifically at the hospitals' trauma center and those receiving the patient initially.

The envisioned scenario can be split up into three phases:

1. *The time from the accident until an ambulance arrives at the site* (Figure 10.3). An alarm call is sent to the authorities:
   - This can be sent automatically or by the patient directly from a device within the patient's own P-PAN.
   - Alternatively, a witness can send the alarm call from a device within her/his P-PAN. The patient's identity and potentially his medical characteristics, if available, may be transmitted together with the call.
   - Localization information will be transmitted.

2. *The initial treatment and transportation to the hospital* (Figure 10.4). An ambulance arrives at the site of the accident and the paramedics initiate treatment:
   - The ambulance commences travel back to the hospital. The patient's P-PAN will interact with the ambulance network.
   - Various devices for diagnosing and monitoring the patient will produce a large amount of data, which, together with the paramedics' verbal comments, high-resolution video, and X-ray pictures will be transmitted to the hospitals' emergency ward and staff.
   - The patients medical records are transmitted from either his own doctor or hospital records to the trauma team and the paramedics in the ambulance.

   Before the ambulance reaches the hospital, the doctor(s) in the hospital's trauma center can evaluate the information available, commence treatment remotely in the ambulance, and continuously monitor the state of the patient. An operating room might be prepared and other specialists might be called upon to participate.

3. *Arrival at hospital and transfer to the receiving personnel* (Figure 10.5). Proper treatment is initiated on the basis of all the information transmitted to the hospital since the time of the incident.

Figure 10.3 Phase 1: persons performing first aid before arrival of ambulance.

Figure 10.4 Phase 2: treatment in an ambulance.

Figure 10.5 Phase 3: the patient arrives at the hospital.

Requirements and Characteristics

Service discovery and interaction of P-PANs belonging to different people are obviously a key issue, but also (and to a very high degree) security will be an issue. There is a very high requirement on speed, necessitating efficient and unsupervised interaction between medical apparatus and P-PANs. The ambulance will potentially traverse many kinds of WANs (UMTS, satellite) at high speed. Localization will also be an essential parameter.

The following list briefly points out the main communication requirements:

- Ambulance traversing heterogeneous networks at high speed;
- Need for localization information for patient, doctors, and ambulance;
- Continuous information exchange: hospital ← → ambulance;
- Ambulance PAN includes devices to measure ECG, blood oxygen, administered drugs, auxiliary condition facts, and so forth;
- Patient information needed in the ambulance: electronic patient journal (EPJ), patient's medical history, and auxiliary medical support;

- If remote treatment is commenced by trauma center doctors, high-resolution video is required.

Emergency scenario characteristics include:

- Adaptation to dynamically changing network environments;
- High requirements: bandwidth, robustness, handover, and security;
- Complex ambulance area network (AAN);
- Range of medical devices on board;
- Temporal inclusion/exclusion of devices in ambulance PAN;
- Air interfaces/protocols;
- A potentially high number of individual P-PANs need to interact.

10.3 Smart Shopping Case

This section describes events related to a user who seamlessly uses personal devices for everyday tasks such as shopping, interacting, and consuming content. The scenarios in this case aim to describe how carrying out the tasks becomes easier and more enjoyable using different types of services through a PAN.

The core device of the user's PAN is the user's smart mobile phone, which is a multiaccess mobile device through which the user can obtain services from different network infrastructures. The multiaccess capability is important because customers should be able access services independently from the underlying wireless access technologies.

The PAN device is attached to the owner and thus is with the owner as he or she moves from one location to another. In a *consumption scenario* the customer's PAN can be augmented by adding to it an MP3 player, mobile mass storage,[1] a wireless headset, a home stereo, product sensors, and their friend's devices. It is easy to envision a PAN further enhanced to include devices such as home sensors, a TV, a digital camera, and so forth. The infrastructure including the different devices is depicted in Figure 10.6.

Usage scenarios within smart shopping can be realized by the following story of shoppers going to and from a shopping mall.

Feed supply info. Paul and Mary are about go to the shopping mall. Mary browses through her list of favorite recipes displayed on her fridge door and selects one. Mary downloads the recipe on her smart mobile device and is ready

---

1. This device is considered one of the main motivators for PAN communications (multipart terminal). Having a mobile mass storage device means that there is no need for annoying synchronization of data on the different devices, because they can all access *the same data*, at home or anywhere else. To prevent data loss, a backup at home is necessary, but synchronization is not.

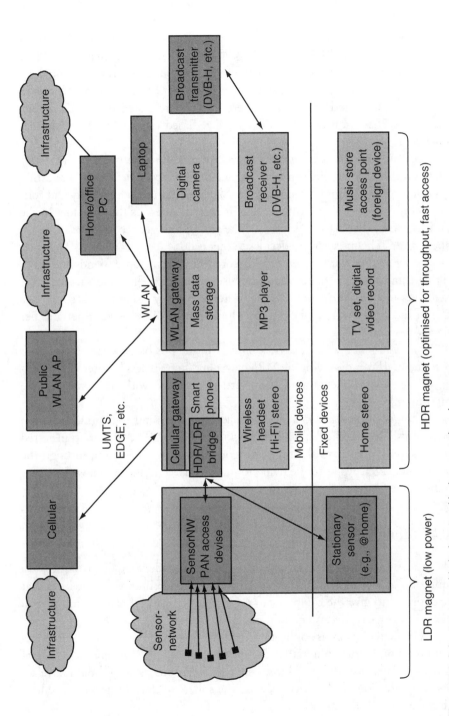

Figure 10.6 Infrastructure with devices used in the smart shopping case.

to leave. Before leaving, sensors at home feed info about current food stock into Mary's smart mobile phone.

The mobile application on Mary's smart phone creates a shopping list of the missing items, which is also used as a preference list for receiving ads. Mary further updates her preference list with other items she is interested in buying.

Receive ads. When entering the mall, Mary switches her mobile device to shopping mode. Food and music advertisements are stored in Mary's mobile mass storage based on Mary's preference list. Based on the ads, Mary decides which stores to visit.

Get recommendations and buy music. After some grocery shopping, Mary enters a music store to buy some music. She finds a record that looks interesting and wants to know more about the band. She joins a music community and connects with other people that share the same music interests that she does so she can learn what they have to say about the record and get recommendations on other bands. The music Mary decides to buy is downloaded to her MP3 player.

Share pictures with friends. At the store Mary meets some friends and starts talking about the great places they have visited during their recent vacation. She exchanges pictures and video clips with her friends by using the user interface (UI) of the smart mobile phone to send and receive files between her and her friends' devices.

Listen to music on the bus back home. On the bus back home Mary listens to her new MP3 audios with the MP3 player and her wireless headset while reading the metadata (background about artist) provided with MP3 files on the smart mobile phone.

Seamlessly enjoy content at home. Back home she wants to continue to listen to the music in the kitchen while unpacking her bags. The files are transferred from the MP3 player to the home stereo, which plays the music through the loudspeakers. She walks to the living room to relax. The audio follows her, now to the home stereo in the living room.

10.4 Student Case

The student case describes several situations of a student's daily activities at a campus. The campus infrastructure provides many technological facilities that are designed to give the campus staff a great deal of helpful features in order to make the everyday life easier and more convenient.

The field of campus applications and services are not limited by purely educational boundaries, but rather extend to include more general activities of daily life. Thus, a student (or more generally a campus person) should have access to applications and services beyond the sphere of the education, like entertainment and social services.

10.4.1 Scenario Description

A senior student, Sam, goes to campus with a high-featured mobile phone, an MP3 player, and a wireless headset. These devices form Sam's P-PAN. Sam has a personal record of the campus database, which gives him access to many campus facilities and services, like the library and the athletic center. Personal student records can be exploited in several ways, as Sam will also get permission to access the campus wireless or wired LAN, he will be able to connect to the Internet and to get access to many other innovative applications that are offered by the campus.

Sam has his own office, as well, which has been given to him for the purpose of writing his thesis. His office contains a desktop PC, a printer, and a scanner. His office is connected with the laboratory, where the experiments related to his thesis are conducted, through the campus WLAN. The above devices form another PAN (the office PAN), and they have a previous correlation with each other. Also, the office PAN is configured to recognize Sam's P-PAN.

What's more, Sam also owns a home PAN, which includes a desktop PC, a printer, a scanner, a camera, a DVD player, a home audio system, and a huge screen. These devices also have a preestablished trust with Sam's P-PAN (and therefore, with his office PAN). Home and office PAN can be considered as *home* and *office cluster*, respectively.

It is assumed that all devices are PAN-enabled; that is, they contain the appropriate hardware and software to enable them to automatically communicate with each other when active (implicit within this assumption is the use of a PAN air interface with appropriate control channels for control of a local PAN).

In this scenario, besides being involved in several social activities around the campus, Sam wishes to carry out an experiment using appropriate computing resources and sensors that are located in his immediate work area at the university. The devices that are at his disposal in the laboratory, amongst others, would include computer and computer peripherals (keyboard, mouse, screen). The device that is the default master is the user's mobile phone.

The experiment is a nutritional experiment that alters the blood pressure monitors of mice. A sensor device is planted into the bodies of 20 mice. The sensor also measures the blood pressure and broadcasts periodic measurements. The measurements are transmitted to a receiver that is inside the laboratory, and then they are relayed, through the receiver, to Sam's preferred device (which can be his PDA, his mobile phone, or his home or office PC).

Sam is also involved in an advanced and modern skill program that takes advantage of university activities and courses to help students be better prepared in their current environment and also in their future working life. This program helps to define skills that are associated to the student's objectives according to his rhythm of education.

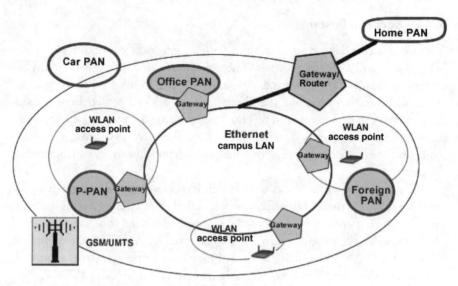

Figure 10.7 Campus network.

The university campus provides many access interfaces to local and global services and networks. Figure 10.7 illustrates the different facilitating networks that the campus provides to its students and staff.

10.5   Mobile Gaming

In this case the gamer will be accessing other PNs. We do not assume any mobility since the gamer will be stationary while playing.

We can foresee the following two scenarios:

1. Gamer joins a multiplayer game mode (on-line).
2. Gamer initiates a multiplayer game mode (peer-to-peer, off-line).

The section will begin with addressing the target market, followed by describing the gaming scenario. Finally, the use cases will be discussed, as will the requirements for accomplishing the described case.

10.5.1   Scenario Description

Gamer Joins Multiplayer Game Mode (On-Line)

John is walking downtown and wants to have some fun. He uses his PN terminal to check for a virtual network game shop. He finds one, but it being his first time in this city, he downloads a map of the area. He eventually finds the store and rents for 1 hour a sensor suit, a gun, and glasses. The proprietor of the shop tells him that he will enter a game that began 10 minutes ago and which involves seven other gamers already connected to the gaming network. He attaches a DVD player to his belt that will display on his glasses, through wireless communication, video information of the games (such as maps and landscapes). The retailer tells him about his goal for the game. He heads to the gaming area. John stands on a sensor carpet, called a game board, which reads his motions. The carpet is mounted on wheels that are easy or hard to move according to the scenery in the game.

John is represented in the game by an avatar named VJohn (Virtual John). John's motions are grabbed by the sensor network he wears and transmitted to the game base station that computes the attitude of his body (arms, legs, head direction). A localization device enables the game base station to describe his position in the gaming area. The base station processes the player's information and determines the position and attitude of the avatar in the game. The gamer processor reckons the viewing angle of the avatar and transmits the scenery from the DVD accordingly. For the other gamers, VJohn's position is overlaid on the scenery. VJohn may be seen or not according to their positions in the game.

As John (virtually) gets closer to another gamer, he automatically receives some information about him (number of lives left, weapon status, hints about his quest). He can see that VSteve is hostile and very well trained. He prefers not to fight this time and escapes. While running he gets lost and he needs to ask for his position; he also asks for other gamers' positions. He sees that there is a more deadly gun close to his position. He heads there. On his way, he meets VAndy with whom he begins to fight. During the fight, Andy is injured and his remaining points decrease. This information is transmitted to John (LDR PN). John now takes the weapon. His profile is updated for the other users. John finishes his game. He would like to go for another game but it is time to go home. The retailer tells him that he can rent the equipment to play at home. "All the equipment, including the gaming board is WLAN compliant. You just have to be connected to our Web page, enter your user name and password, and you will get instructions on how to join our network virtual gaming area. Enjoy!"

Gamer Joins Multiplayer Game Mode (Off-Line)

The gamer visits his friends and they build a peer-to-peer network between their P-PANs. The glasses, microphones, and headphones are used from John's other personal devices. John and his friends connect using either UWB and or WLAN as the legacy systems. The game is set up and played similarly as during the online scenario.

10.6 Distributed Work Cases

A distributed work case refers to applications for people who are mobile, such as journalists, tourists, and truck drivers, and who utilize any of the devices available to conduct their tasks. The nomadic user needs to perform tasks remotely through their PN. We illustrate scenarios involving washing machine repair, virtual truck, journalism, and printing to elucidate the theme of distributed work case.

10.6.1 Washing Machine Repair Service

This case is concerned with the household environment. The P-PAN at home includes the appliances within the household and other electronic devices like the PC or PDA. This case specifically considers a washing machine, which is controlled remotely by the homeowner's mobile phone, PDA, or laptop, but is now defective and requires servicing. A repair service enters the home and needs to use some services of the P-PAN, so a security mechanism must be used to give the repair service a restricted access to the P-PAN. The repair service connects to the washing machine and also to the company network over the Internet access of the household. This is illustrated in Figure 10.8.

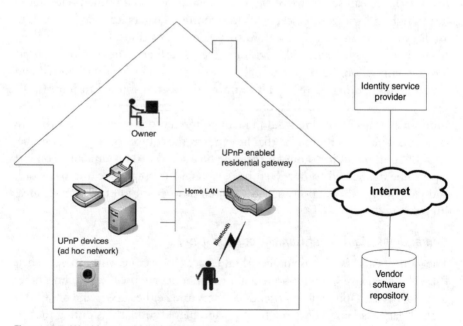

Figure 10.8 Washing machine repair service.

The homeowner can control his domestic appliances with his laptop or PDA while he is within his house or outside. Depending on the type of appliance, he can start or stop them or set some control parameters. For example, the inhabitant can set the temperature and the program of his washing machine and start it.

The homeowner also has the possibility of getting status information about the devices. The washing machine can inform the user when the washing process is finished. In the case of a dysfunction, the appliance can also send an error message to the homeowner.

The error message can be forwarded to the repair service, as shown in Figure 10.9. The error message indicates the need to send a repair technician to the malfunctioning device. This repairman needs to access the P-PAN of the household for accessing the washing machine with his PDA. The homeowner will give him restricted access to the necessary resources (e.g., the washing machine, Internet access). The service man is now able to check the washing machine physically and with his PDA.

One reason for the malfunction can be old and buggy software. In this case the service technician will connect to the software repository of his company network in order to download the new software for the specific device. He needs to access the company cluster to form a PN with his PDA and the software repository. After access is granted, he downloads the software update from the company server and installs it on the washing machine.

10.6.2 Virtual Home Truck

Transportation and logistics represent a major business industry employing millions of truck drivers. Each day, these people spend hours in their vehicles, either driving, waiting, or sleeping; they are often away from home for many days at a time. Offering these people the ability to stay in touch with their family (by creating a virtual home environment), with their company and clients, or their fellow truck drivers, could have great commercial potential.

This vision can be realized through the concept of virtual home trucks. Consider a truck equipped with a mobile phone, broadband Internet access [6], TFT display, headset, speakers, and laptop, all of which form a cluster of cooperating devices. When finished with work, a truck driver could set up an Internet connection to his home. At home, a cluster of cooperating cameras, speakers, and headsets, provides the truck driver with a virtual home environment. Through this environment, he can virtually walk around at home, seeing his family, talking with them. He can play games with his children or watch a movie together with his family.

When driving, the truck driver can listen to his digital music collection by streaming it from a server in his network at home. Meanwhile, his point of

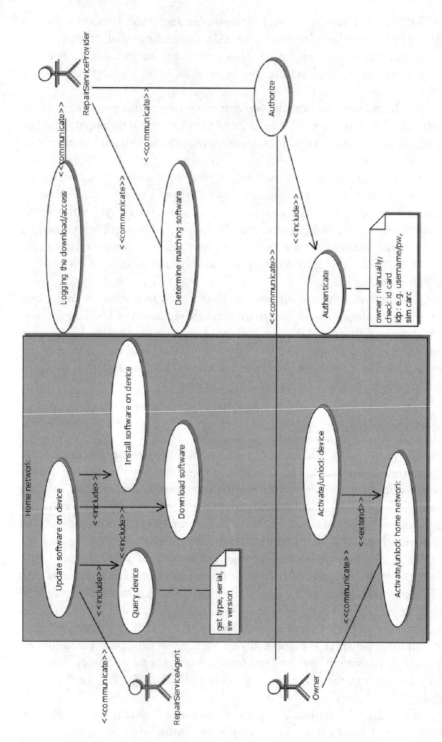

Figure 10.9 Use case diagram.

attachment to the interconnecting infrastructure can change, making mobility management a necessity. When the truck driver stops at a parking area, he can read his e-mail, he can search the parking lot for colleagues, he can play a multihop game with other truck drivers.

When the truck driver arrives at a client, his PAN can connect to the client's company network and download the necessary documents. The documents can be digitally signed, handed over to the client and a copy can be uploaded to the truck drivers company, reducing the administrative burden.

10.6.3 Distributed Work Case for Journalists

The overall aim of the distributed work case is to analyze and identify how an extended use of ICT can influence professional working environments regarding work processes and organizational forms. Broadcast services, including the news services, are heavily based on high-quality video/audio components, and services are generated in a complex interaction between different professionals in the field and also their interaction with the home company. Another important aspect is that the services are generated in highly mobile environments, and as a consequence the bandwidth requirements to PAN and PN networks in this case are very high. Therefore, the design of efficient communication and interaction paths in this context is quite complex.

The main purpose of this distributed work case is to analyze the professional working environment for advanced users of PA/PANs by studying the work processes within the field of journalism in general and in particular regarding news journalists and others persons/technicians involved in the task of gathering, producing, and finalizing a video or audio news story from anywhere in the world. The main focus will be related to analyzing and describing how extended mobility in specific organizations can be supported by personal and personal area networks regarding work processes and organizational structures, and furthermore to identify the drivers and barriers for the introduction of new technologies and services.

Basically, this means that the distributed work case will consists of several PN/PANs (those of journalists, camera operators, sound operators, and news team members), which all individually consists of several devices that needs to interconnect and exchange data within a specific PAN but also with other PANs through different PNs using different heterogeneous networks.

10.6.4 Remote Printing Service

This case depicts a scenario where a person can use his handheld device (e.g., mobile phone) to print pictures on his home printer while he is travelling far away from his home. This service is called *remote printing*.

Jack takes a picture using his mobile phone when he is travelling around the city, and he wants to show this amazing picture to his family right away. He finds a public access gateway (AGW) nearby, which discovers Jack's mobile phone via Bluetooth. Jack sends a request asking for services from the public-AGW. If the mobile phone has not registered on the public AGW yet, the gateway will register for it and issue service discovery request to Jack's personal AGW at home. The personal AGW will discover services available at the home network and send a service list back to the public AGW. The public AGW will then propagate this service list to Jack's mobile phone, and also store this service list combined with some information related to Jack (e.g., his mobile phone's MAC address and the IP of his personal AGW) in its database for future use. Now, Jack can select the printing service from the service list on his mobile's screen. Then he is informed to select the file that he wants to print. Finally the picture will be delivered to the personal AGW, which will handle printing issues.

10.7 Conclusions

Personal networks enable users to experience greater accessibility to services, which ultimately accounts for substantial value added for content providers. The socioeconomic benefits derived from this network are apparent, as is illustrated in the health care scenario. The applications foreseen are enhancements of some existing market cases, such as mobile gaming. Moreover, such as in the distributed work arena, users have the option to use multiple devices to access the same service, which allows the user to choose the level of presence they like to experience. Furthermore, PN ushers in dynamism to the paradigm of doing business.

The business models in this environment will be influenced greatly since financial flow would be diverted from a centralized mobile business infrastructure to a decentralized infrastructure. Billing and charging will be issues that require different handling from the conventional methods; use of services must be charged and then billed in a PN environment where interconnections are much more complex and the network handover is more dynamic. Profile management would require essentially three components: profile server, location server, and content/context adaptor. A user's location will be determined and the service can be provided, location specific, such as in the case of an emergency, or remote printing, and so forth. All the cases envisioned in this network would require new solutions for the network, which would have to be add-ons to the conventional architecture.

References

[1]   Olsen, R. L., et al., "User Centric Service Discovery in Personal Networks," *Proceedings International Symposium WPMC04*, Abano Terme, Italy, September 2004.

[2]   http://www.aarhus.dk/aa/portal/borger/s_nyheder/indhold?articleId=19550&_refresh=true.

[3]   http://www.it-raad.dk/taenketanke/ pervasivehealth/vision_arbejdsrutiner.doc.

[4]   http://cph.ing.dk/tema/telemedicin/tele.html.

[5]   http://tie.telemed.org/.

[6]   http://www.its.dot.gov/.

[7]   Jiang, B., and H. Olesen, "Agent-Based Personal Network (PN) Service Architecture," *Proceedings International Symposium WPMC'04*, Abano Terme, Italy, September 2004.

[8]   Prasad, R. R., and M. Monti, "Pricing Policy for Personal Network within Magnet Project," *Proceedings International Symposium WPMC'04*, Abano Terme, Italy, September 2004.

List of Acronyms

AA	Authentication Agent
AAA	Administration, Authorization, and Authentication
ACH	Access Channel
ACK	Acknowledgement
ACL	Asynchronous ConnectionLess
ACL	Access Control List
ACM	Association of Computing Machinery
ACO	Authentication Ciphering Offset
ADSL	Asynchronous Digital Subscriber Loop
AES	Advanced Encryption Standard
AI	Ambient Intelligence
AIFS	Arbitration Interframe Space
AK	Authorization Key
AKA	Authentication and Key Agreement
AODV	Ad-hoc On Demand distance Vector routing
AP	Access Point
API	Application Programming Interface
APS	Asynchronous Power Save
ARP	Address Resolution Protocol
ARQ	Automatic Repeat Query
ASK	Amplitude Shift Keying
AS	Authentication Server
ASB	Active Slave Broadcast
ATM	Asynchronous Transfer Mode
BAN	Body Area Networks
BCH	Broadcast Channel
BER	Bit Error Rate

BKR	Broadcast Key Rotation
BNEP	Bluetooth Network Encapsulation Protocol
BOK	Binary Orthogonal Keying
BPSK	Binary Phase Shift Keying
BRAN	Broadband Radio Access Networks
BRP	Bordercast Resolution Protocol
BS	Base Station
CAN	Community Area Network
CAP	Contention Access Period
CAZAC	Constant Amplitude Zero Autocorrelation
CCK	Complementary Code Keying
CDMA	Code Division Multiple Access
CEPT	Conférence Européenne des Postes et des Télécommunications
CFP	Contention Free Period
CN	Core Network
CoA	Care of Address
COF	Ciphering Offset
CP	Contention Period
CRC	Cyclic Redundancy Check
CSMA/CA	Carrier Sense Multiple Access / Collision Avoidance
CTA	Channel Time Allocations
CTAP	Channel Time Allocation Period
CVSD	Continuously Variable Slope Delta
CW	Contention Window
D-MAC	Directional MAC
DAG	Directed Acyclic Graph
DCF	Distributed Coordination Function
DFWMAC	Digital Foundation Wireless MAC
DLC	Data Link Control
DLL	Data Link Layer
DNAV	Directional NAV
DOA	Direction of Arrival
DoS	Denial of Service
DQPSK	Differential Quaternary Phase Shift Keying
DSDV	Destination Sequenced Distance Vector
DSPS	Device Synchronized Power Save
DSR	Dynamic Source Routing
DSSS	Direct Sequence Spread Spectrum
DVB	Digital Video Broadcating
EAP	Extensible Authentication Protocol
EAPOL	Extensible Application Protocol Over LAN
ECG	ElectroCardioGram

ECWT	European Conference on Wireless Technologies
EDCF	Enhanced Distributed Coordination Function
EIFS	Extended Interframe Space
EIRP	Equivalent Isotropic Radiated Power
ETSI	European Telecommunication Standard Institute
FA	Foreign Agent
FAMA	Floor Acquisition Multiple Access
FCC	Federal Communication Commission
FCS	Frame Check Sequence
FDD	Frequency Division Duplex
FDMA	Frequency Division Multiple Access
FEC	Forward Error Coding
FFD	Full Function Device
FFT	Fast Fourier Transform
FH	Frequency Hopping
FHS	Frequency Hopping Synchronization
FHSS	Frequency Hopping Spread Spectrum
FIFO	First In First Out
FSK	Frequency Shift Keying
FSM	Finite State Machine
FSR	Fisheye State Routing
FTTH	Fiber to the Home
FWA	Fixed Wireless Access
GAP	Generic Access Profile
GFSK	Gaussian Frequency Shift Keying
GGSN	Gateway GPRS Supporting Node
GOEP	Generic Object Exchange Profile
GPRS	General Packet Radio Service
GPS	Global Positioning Systems
GSM	Global System for Mobile
GTS	Guaranteed Time Slot
HA	Home Agent
HC	Hybrid Coordinator
HCI	Host Controller Interface
HCS	Header Check Sequence
HIPERLAN	High Performance Local Area Network
HIPERMAN	High Performance Metropolitan Area Network
HLR	Home Location Register
HRMA	Hop Reservation Multiple Access
HSR	Hierarchical State Routing
IAPP	Inter Access Point Protocol
IARP	Intrazone Routing Protocol

ICT	Information and Communication Technology
ICV	Integrity Check Value
IDEA	International Data Encryption Algorithm
IEEE	Institute of Electrical and Electronics Engineers
IETF	Internet Engineering Task Force
IF	Intermediate Frequency
IFS	Interframe Space
IMEI	International Mobile Equipment Identity
IMSI	International Mobile Subscriber Identity
INS	Intentional Naming System
IP	Internet Protocol
IR	Impulse Radio
ISDN	Integrated Service Digital Network
ISI	Inter Symbol Interference
ISM	Industrial Scientific and Medical
ISO	International Standards Organization
ISP	Internet Service Providers
ITU	International Telecommunication Union
L2CAP	Logical Link and Control Adaptation Protocol
LAI	Location Area Identity
LAN	Local Area Networks
LC	Link Controller
LDR	Low Data Rate
LFSR	Linear Feedback Shift Registers
LLC	Logical Link Control
LM	Link Manager
LMP	Link Manager Protocol
LQI	Link Quality Indication
LR	Low Rate
MAC	Medium Access Control
MACA	Multiple Access Collision Avoidance
MACA-BI	MACA Bi Invitation
MAGNET	My Adaptive Global NETwork
MANET	Mobile Ad-hoc NETwork
MB-OFDM	Multiband OFDM
MBS	Mobile Broadband System
MD5	Message Digest 5
ME	Mobile Equipment
MEM	Micro Electro Mechanical
MIMO	Multiple Input Multiple Output
MMAC	Multimedia Mobile Access Communication
MOP	Million Operations

MPR	Multipacket Reception
MQRS	Multi Queue Room Service
MSC	Message Sequence Chart
NAV	Network Allocation Vector
NDMA	Network Diversity Multiple Access
O-QPSK	Offset Quadrature Phase Shift Keying
OBEX	Object Exchange
OCB	Offset Codebook
OFDM	Orthogonal Frequency Division Multiplexing
OMS	Operation and Maintenance Subsystem
OO	Object Oriented
OS	Operating System
OSA	Open Systems Authentication
OSI	Open Standard Interface
PACWOMAN	Power Aware Communications for WPANs
PAE	Port Access Entity
PAMAS	Power Aware Multi-Access protocol with Signalling
PAN	Personal Area Network
PARO	Power Aware Routing Optimization
PCF	Point Coordination Function
PCM	Pulse Coded Modulation
PCMCIA	Personal Computer Memory Card International Association
PDA	Personal Digital Assistant
PDC	Personal Digital Cellular
PDU	Protocol Data Unit
PEAP	Protected EAP
PIFS	PCF Interframe Space
PIM	Personal Information Management
PIN	Personal Identification Number
PMP	Point to Multipoint
PN	Personal Network
PNC	Piconet Network Coordinator
POS	Personal Operating Space
POTS	Plain Old Telephone Systems
PPP	Point to Point Protocol
PRN	Pseudorandom Number
PRNET	Packet Radio Network
PRNG	Pseudorandom Number Generator
PS	Packet Sensing
PS	Packet Switched
PSB	Parked Slave Broadcast
PSK	Phase Shift Keying

PSPS	Piconet Synchronized Power Save
PSTN	Public Switched Telephone Network
QAM	Quadrature Amplitude Modulation
QoS	Quality of Service
QPSK	Quaternary Phase Shift Keying
RADIUS	Remote Authentication Dial in User Service
RC5	Rivest Cipher 5
RF	Radio Frequency
RFD	Reduced Function Device
RFID	Radio Frequency Identifier
RREP	Route Reply
RREQ	Route REQuest
RSA	Rivest, Shamir, & Adleman (public key encryption technology)
RSN	Robust Security Network
RSS	Received Signal Strength
RTR	Request to Receive
RTS/CTS	Request to Send Clear to Send
SAID	Security Association Identities
SCCAN	Seamless Converged Communication Across Networks
SCH	Synchronization Channel
SCO	Synchronous Connection Oriented
SDMA	Space Division Multiple Acces
SDP	Service Discovery Protocol
SDU	Service Data Unit
SGSN	Serving GPRS Support Node
SIFS	Short Interframe Space
SIG	Special Interest Group
SIM	Subscriber Identity Module
SLP	Service Location Protocol
SM	Spatial Multiplexing
SNR	Signal to Noise Ratio
SOC	System on Chip
SRES	Signed Response
SRP	Secure Remote Password
SSDP	Simple Service Discovery Protocol
TCM	Trellis Coded Modulation
TCP	Transmission Control Protocol
TDD	Time Division Duplex
TDMA	Time Division Multiple Access
TEK	Traffic Encryption Key
TFT	

TG3A	
TKIP	Temporal Key Integrity Protocol
TLS	Transport Layer Security
TMSI	Temporary Mobile Subscriber Identity
TORA	Temporally Ordered Routing Algorithm
TS	Tail Symbols
TTLS	Tunneled TLS
U-NII	Unlicensed National Information Infrastructure
UART	Universal Asynchronous Receiver Transmitter
UMA	Unlicensed Mobile Access
UMTS	Universal Mobile Telecommunications System
USB	Universal Serial Bus
USIM	UMTS Subscriber Identity Module
UWB	Ultra WideBand
VAP	Virtual Access Point
VCO	Voltage Controlled Oscillator
VLR	Visitor Location Register
VLSI	Very Large Scale Integration
VoIP	Voice over IP
VoWLAN	Voice over WLAN
VPN	Virtual Private Network
WAN	Wide Area Network
WCDMA	Wideband CDMA
WEBS	Wireless Embedded Systems
WECA	Wireless Ethernet Compatibility Alliance
WEP	Wired Equivalent Privacy
WLAN	Wireless Local Area Network
WPA	Wi-Fi Protected Access
WRAP	Wireless Robust Authenticated Protocol
WRP	Wireless Routing Protocol
WWAN	Wireless Wide Area Network
WWRF	Wireless World Research Forum
ZRP	Zone Routing Protocol

About the Authors

Ramjee Prasad received his B.Sc. (eng.) from the Bihar Institute of Technology, Sindri, India, and his M.Sc. (eng.) and Ph.D. from the Birla Institute of Technology (BIT), Ranchi, India, in 1968, 1970, and 1979, respectively.

He joined BIT as a senior research fellow in 1970 and became an associate professor in 1980. While he was with BIT, he supervised a number of research projects in the area of microwave and plasma engineering. From 1983 to 1988, he was with the University of Dar es Salaam (UDSM), Tanzania, where he became a professor of telecommunications in the Department of Electrical Engineering in 1986. At UDSM, he was responsible for the collaborative project Satellite Communications for Rural Zones with Eindhoven University of Technology, the Netherlands. From February 1988 through May 1999, he was with the Telecommunications and Traffic Control Systems Group at Delft University of Technology (DUT), where he was actively involved in the area of wireless personal and multimedia communications (WPMC). He was the founding head and program director of the Center for Wireless and Personal Communications (CWPC) of International Research Center for Telecommunications—Transmission and Radar (IRCTR). Since June 1999, Dr. Prasad has been with Aalborg University, as the codirector of the Center for PersonKommunikation (CPK) until 2002, and since January 2003 as the research director of the Department of Communications Technology. He holds the chair of wireless information and multimedia communications. He was involved in the European ACTS project FRAMES (Future Radio Wideband Multiple Access Systems) as a DUT project leader. He is a project leader of several international, industrially funded projects. He is the project coordinator of the European sixth framework integrated project "My Personal Adaptive Global NET (MAGNET)." He has published more than 500 technical papers, contributed to several books, and has authored, coauthored, and edited 16 books: *CDMA for Wireless Personal Com-*

munications; Universal Wireless Personal Communications, Wideband CDMA for Third Generation Mobile Communications; OFDM for Wireless Multimedia Communications; Third Generation Mobile Communication Systems; WCDMA: Towards IP Mobility and Mobile Internet; Towards a Global 3G System: Advanced Mobile Communications in Europe, Volumes 1 & 2; IP/ATM Mobile Satellite Networks; Simulation and Software Radio for Mobile Communications; Wireless IP and Building the Mobile Internet; WLANs and WPANs towards 4G Wireless; Technology Trends in Wireless Communications; Multicarrier Techniques for 4G Mobile Communications; OFDM for Wireless Communication Systems; and *Applied Satellite Navigation Using GPS, GALILEO, and Augmentation Systems,* all published by Artech House. His current research interests lie in wireless networks, packet communications, multiple-access protocols, advanced radio techniques, and multimedia communications.

Dr. Prasad has served as a member of the advisory and program committees of several IEEE international conferences. He has also presented keynote speeches, and delivered papers and tutorials on WPMC at various universities, technical institutions, and IEEE conferences. He was also a member of the European Cooperation in the Scientific and Technical Research (COST-231) project dealing with the evolution of land mobile radio (including personal) communications as an expert for the Netherlands, and he was a member of the COST-259 project. He was the founder and chairman of the IEEE Vehicular Technology/Communications Society Joint Chapter, Benelux Section, and is now the honorary chairman. In addition, Dr. Prasad is the founder of the IEEE Symposium on Communications and Vehicular Technology (SCVT) in the Benelux, and he was the symposium chairman of SCVT'93.

In addition, Dr. Prasad is the coordinating editor and editor-in-chief of the *Kluwer International Journal on Wireless Personal Communications* and a member of the editorial board of other international journals, including the *IEEE Communications Magazine* and *IEE Electronics Communication Engineering Journal.* He was the technical program chairman of the PIMRC'94 International Symposium held in The Hague, the Netherlands, September 19–23, 1994, and also of the Third Communication Theory Mini-Conference in Conjunction with GLOBECOM'94, held in San Francisco, California, November 27–30, 1994. He was the conference chairman of the fiftieth IEEE Vehicular Technology Conference and the steering committee chairman of the second International Symposium WPMC, both held in Amsterdam, the Netherlands, September 19–23, 1999. He is the general chairman of WPMC'01, which was held in Aalborg, Denmark, September 9–12, 2001.

Dr. Prasad is also the founding chairman of the European Center of Excellence in Telecommunications, known as HERMES. He is a fellow of IEE, a fellow of IETE, a senior member of IEEE, a member of the Netherlands Electronics and Radio Society (NERG), and a member of IDA (Engineering Society in Denmark).

Luc Deneire received an engineering degree in electronics from University of Liege, Belgium, in 1988, an engineering degree in telecommunications from University of Louvain-La-Neuve in 1994, and a Ph.D. degree in signal processing from Eurecom, Sophia-Antipolis, France in 1998. During this time, he was a Marie Curie Fellow grant holder.

From 1999 to 2002, he was a senior researcher at IMEC, Leuven, Belgium, the largest European independent research institute in microelectronics, and recently joined the Signals and Systems (13S) of the University of Nice, Sophia-Antipolis, France.

He is working on signal processing algorithms involved in wireless communications, specifically for third generation mobile networks, wireless LANs, and wireless personal area networks. His main interests are blind and semi-blind equalization and channel estimation, modulation theory, multiple access schemes, smart antennas, and link adaptation. He is the author of more than 50 conference and journal papers and two book chapters.

In IMEC, he was responsible for wireless personal area networks and has initiated and led the PACWOMAN (Power Aware Communications for Wireless OptiMised personal Area Network) European project.

Index

The Artech House Universal Personal Communications Series

Ramjee Prasad, Series Editor

Wireless Communications Security, Hideki Imai, Mohammad Ghulam Rahman, and Kazukuni Kobara

Wireless IP and Building the Mobile Internet, Sudhir Dixit and Ramjee Prasad, editors

WLAN Systems and Wireless IP for Next Generation Communications, Neeli Prasad and Anand Prasad, editors

WLANs and WPANs towards 4G Wireless, Ramjee Prasad and Luis Muñoz

For further information on these and other Artech House titles, including previously considered out-of-print books now available through our In-Print-Forever® (IPF®) program, contact:

Artech House	Artech House
685 Canton Street	46 Gillingham Street
Norwood, MA 02062	London SW1V 1AH UK
Phone: 781-769-9750	Phone: +44 (0)20 7596-8750
Fax: 781-769-6334	Fax: +44 (0)20 7630-0166
e-mail: artech@artechhouse.com	e-mail: artech-uk@artechhouse.com

Find us on the World Wide Web at: www.artechhouse.com